THE FUTURE OF THE GULF:
Politics and Oil in the 1990s

For Verity with much love

THE FUTURE OF THE GULF:
Politics and Oil in the 1990s

Philip Robins

with a chapter by Jonathan Stern

The Royal Institute of International Affairs
Dartmouth

British Library Cataloguing in Publication Data

Robins, Philip
The future of the Gulf: politics and oil in the 1990s
(The Energy and Environmental Programme; 25)
1. Western world petroleum supply. Implications of political events in petroleum exporting countries in Persian Gulf.
I. Title II. Stern, Jonathan P., 1951- III. Roya Institute of International Affairs IV. Series
382'.42282'0911821

ISBN 1 85521 011 8

Published by Dartmouth Publishing Company, Gower House, Croft Road, Aldershot, Hants GU11 3HR, England

Gower Publishing Company, Old Post Road, Brookfield, Vermont 05036, USA

ISBN 1 85521 011 8

Printed and Bound by Athenæum Press
Newcastle upon Tyne

Contents

Preface

This book is the first major piece of research to be published under the auspices of the Royal Institute of International Affairs Middle East Programme, after its relaunching in 1987 with Dr Robins as Research Fellow.

Dr Robins' well-researched conclusions are based not only on an academic study of the area but also on many conversations with persons involved in the decision-making process in the Gulf and further afield.

Though this book stands on its own, it is to be seen as part of a programme of research and publication beginning with the publication in December 1987 of *Egypt: Internal Challenge and Regional Stability*. Work has already started on the next research project entitled *Turkey and the Middle East*. The themes of this study are the effect of such Middle East phenomena as the Kurds and Islamic Fundamentalism on the domestic policies of the country and Turkey's political and economic relations with its neighbours.

In this way a continuing programme of research will be undertaken which will be topical and relevant to those who need to be aware of developments in the Middle East.

Sir John Moberly
January 1989

Acknowledgments

The completion of such a far-ranging project in the relatively short period of 15 months was made possible due to the help and advice of many people. Many of those must remain anonymous as it was on this basis that they so generously and frankly gave of their views. In particular I would thank those who participated in the four study-groups which were convened, and those who gave of their time and hospitality on my frequent research trips.

Closer to home, I would like to thank the staff of Chatham House for providing such a congenial and stimulating context in which to work. The hardworking and cheerful members of the Library and Press Library merit special mention for their considerable help, effectively and happily rendered.

Of my colleagues on the Middle East Programme, Sir John Moberly supervised the project, read the draft and gave wise counsel throughout. I am indebted to the Head of the Programme, Jonathan Stern, in two ways. First, for providing the chapter on oil, and second for arranging the funding of the project. Rosina Pullman, ably assisted by Helen Tomkys and Anne Martin, was responsible for administering the project so efficiently and for keeping the study to its timetable. I am also grateful to Judith Nichol for research assistance on chapter 5. My thanks also go to Margaret Cornell, who edited the manuscript.

Special mention must be made of the National Institute for Research Advancement (NIRA) in Tokyo, who so generously funded the project. Without their assistance, the study would not have been undertaken.

Though the list is long of those to whom gratitude is deserving, the views and interpretations presented are those of the author, and not necessarily those of the members either of the supporting organisations or of institutions that sponsor the Royal Institute of International Affairs.

Foreword

The end of the war has inevitably given a new perspective to a study of the Gulf. The timing of the ceasefire was also significant as far as this study was concerned, since it seeks to analyse the contemporary position in the wake of hostilities. From a strong foundation of understanding of the present, views on the next decade have been taken. Writing in the new context of a cessation of hostilities has proved no difficulty in finding fresh insights. This report is thus one of the first to be produced which looks forward to the 1990s in the aftermath of the conflict.

The main focus has been a general one. Because of the interdependence of the Gulf region, the study has set out to examine all the littoral states. The major regional powers, Iran and Iraq, have been considered individually in some detail. Attention has also been paid to the six members of the Gulf Co-operation Council, and indeed to that multilateral organisation itself. Two main themes have been adopted. Firstly, there has been an attempt to survey the internal politics of the Gulf states, with a particular view to judging the stability of the ruling regimes. Secondly, foreign policy and foreign relations have been considered. While Iran and Iraq have been examined systematically in terms of these themes, it was considered rather unsatisfactory to do the same for all the Gulf states. They have, therefore, been analysed together in a way which it is hoped will be more illuminating and facilitate comparative analysis.

A chapter on the superpowers has been included despite the easing of tension in the Gulf itself. An end to the hostilities does not change the structural interests which the superpowers have in the area. Rather, it was the Gulf conflict which showed so clearly the depth and extent of their interests.

No assessment of the future or indeed the present situation in the Gulf area would be complete without a consideration of the oil dimension. This is particularly true in looking at the next decade. The chapter has been cautious in predicting the future, pointing out the pitfalls of previous conventional wisdom, yet pointing out that if war has been the main challenge to the area in the 1980s, recession born of flat oil revenues could be the challenge of the 1990s.

Much emphasis has been placed on the use of primary sources with visits to all the members of the GCC, plus Iraq, the United States and Japan. Valuable contact has been make with foreign diplomats and decision-makers and businessmen, as well as with local intellectuals and other influential figures. Much of the data on which the analysis has been based has emerged as a result of what have been by and large confidential discussions and interviews with such people. Regrettably, therefore, it has not been possible to refer directly to our sources.

Summary

The geographical location of Iran means that it will physically continue to dominate the Gulf area. Likewise the size and growth of the population will mean that, in spite of the short term weakness sustained as a result of the war, Iran remains important both as a market and as a regional actor. Since the early days of the revolution there has been a considerable degree both of stability and of elite continuity in Iran. This has been helped by the virtually complete absence of any credible, orgainised opposition to the ruling regime. The present leadership enjoys considerable revolutionary legitimacy. This all bodes well for continued stability, even after Ayatollah Khomeini dies. Recognition at all levels of the unpopularity of the war will ensure that the fighting does not start up again. In foreign affairs, there has been a coherent attempt to end the country's isolation. Reconstruction of the civilian economy and the re-arming of the military will take fiscal priority in the future.

The main internal challenges to the state and regime in Iraq during the war have receded. The Kurdish opposition groups have been thrown on the defensive, while the Shi'a threat has proved to have been exaggerated. The importance and resilience of the Ba'th Party are often underplayed. It cannot be assumed, however, that the death of Saddam Husain would precipitate a collapse of the regime. Economic liberalisation looks set to continue, as does incremental progress in political reforms, though within the context of tight central control. Despite the country's strong showing at the end of the war, there is a tremendous popular will for peace. Nevertheless Baghdad will continue to be preoccupied with the threat from the east. As a result, the Iraqi leaders will not want to undermine the ruling regimes in the Gulf states providing they do not change their policies towards Iraq.

The fact that members of the Gulf Co-operation Council want it to continue to function, even though the Gulf conflict appears to be over, suggests perceived utility, not least in helping to ensure the perpetuation of the political order in the member states. The internal threat to those states from their Shi'a communities seems likely to recede in the aftermath of the war. Economic problems and a reduction of resources set against a backdrop of growing populations and burgeoning expectations may give rise to domestic instability in some member states in the 1990s. The immediate challenge is to restore an equilibrium of power in the Gulf which involves both Iran and Iraq in constructive roles.

The Gulf is an area where the interests of both superpowers are involved. The geographical proximity of Iran and the Soviet Union and the former's repeated historical experience of the overbearing nature of the superpower to the north strictly limit the potential for close bilateral relations. The relationship between Moscow and Baghdad has been founded on mutually perceived utility, rather than anything more profound. Iran's relationship with the US is currently at a low level. Ultimately, though, there is greater potential for good relations with Washington than with Moscow. While the new constructive basis of Soviet policy apparently offers the chance of less friction between the superpowers in the Gulf, alarm in Washington over potential diplomatic successes as a restoration of relations with Saudi Arabia, allied with possible renewed superpower jockeying to cultivate Iran, could herald a new chapter of competition.

The concentration of proven oil reserves in the Gulf means that, even during a period of weak demand and low prices, the potential for a few regional exporters to dominate the world oil market will remain on the agenda for the 1990s. There will be continuing attempts on the part of both Iran and Iraq to expand exports as a way of financing their reconstruction efforts. Downward pressure on oil appears likely therefore to continue in the absence of a strong increase in world oil demand or a major military conflagration in the Gulf. However, the Iran-Iraq war showed that even a military conflict is less likely fundamentally to disturb world oil markets, given the diversification of supply routes away from the Gulf waterway, and that regimes, irrespective of ideology, require a minimum level of revenues.

In conclusion, the Gulf will continue to be an important sub-system of the global international community throughout the 1990s due to the concentration of oil reserves; the active interests of the superpowers; and the uncertainty of regional power relations. At the end of the 1980s the prospects for further inter-state conflict in the area appear to be low. However, a reduction in tension in the Gulf in the 1990s compared with the previous decade, could still include domestic challenges to some of the littoral states, perhaps exacerbated by falling revenues against a backdrop of rising populations and economic expectations. The management of resource allocation could emerge as a critical test of the political acumen of the regimes in power.

Map 1. The Middle East

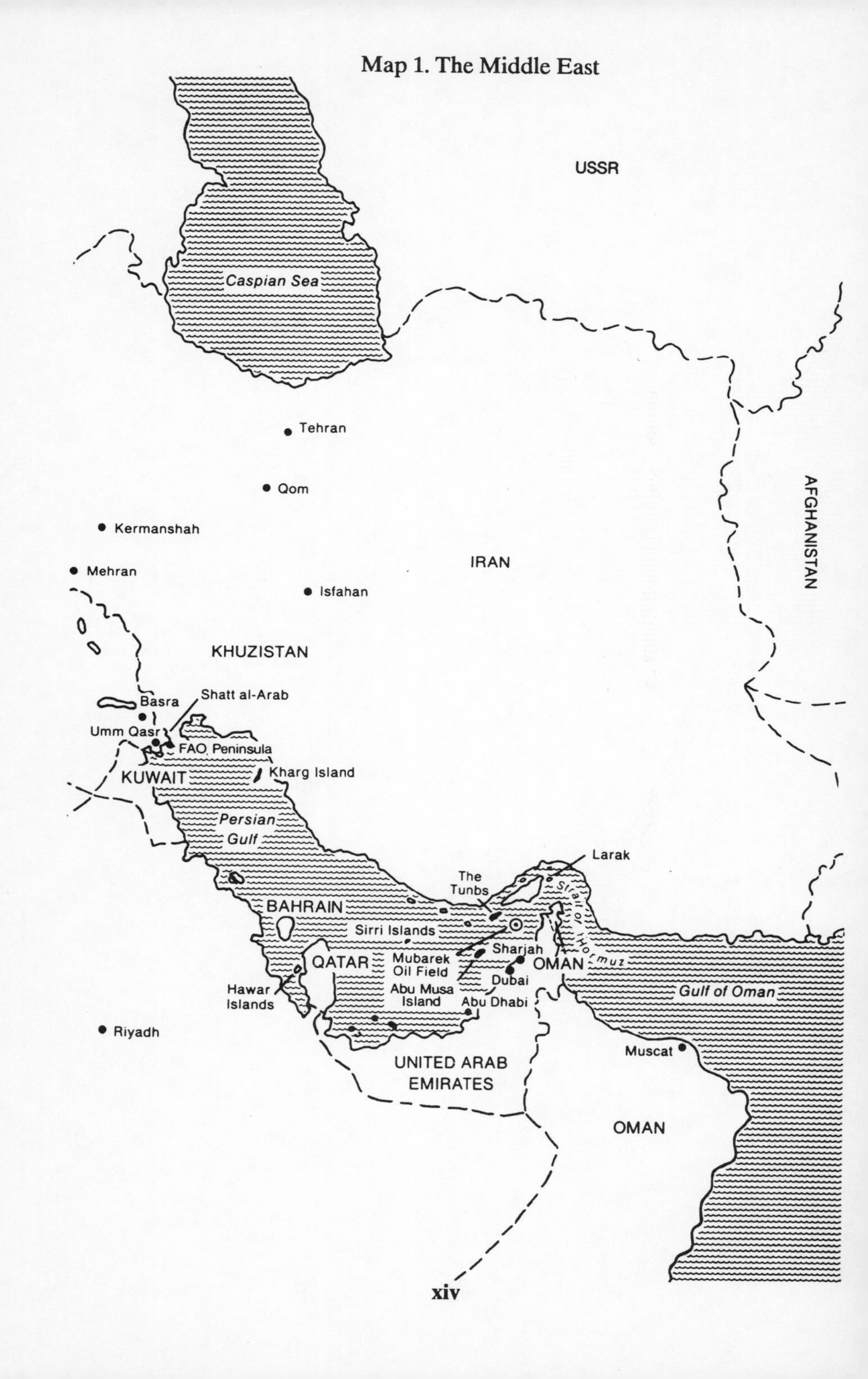

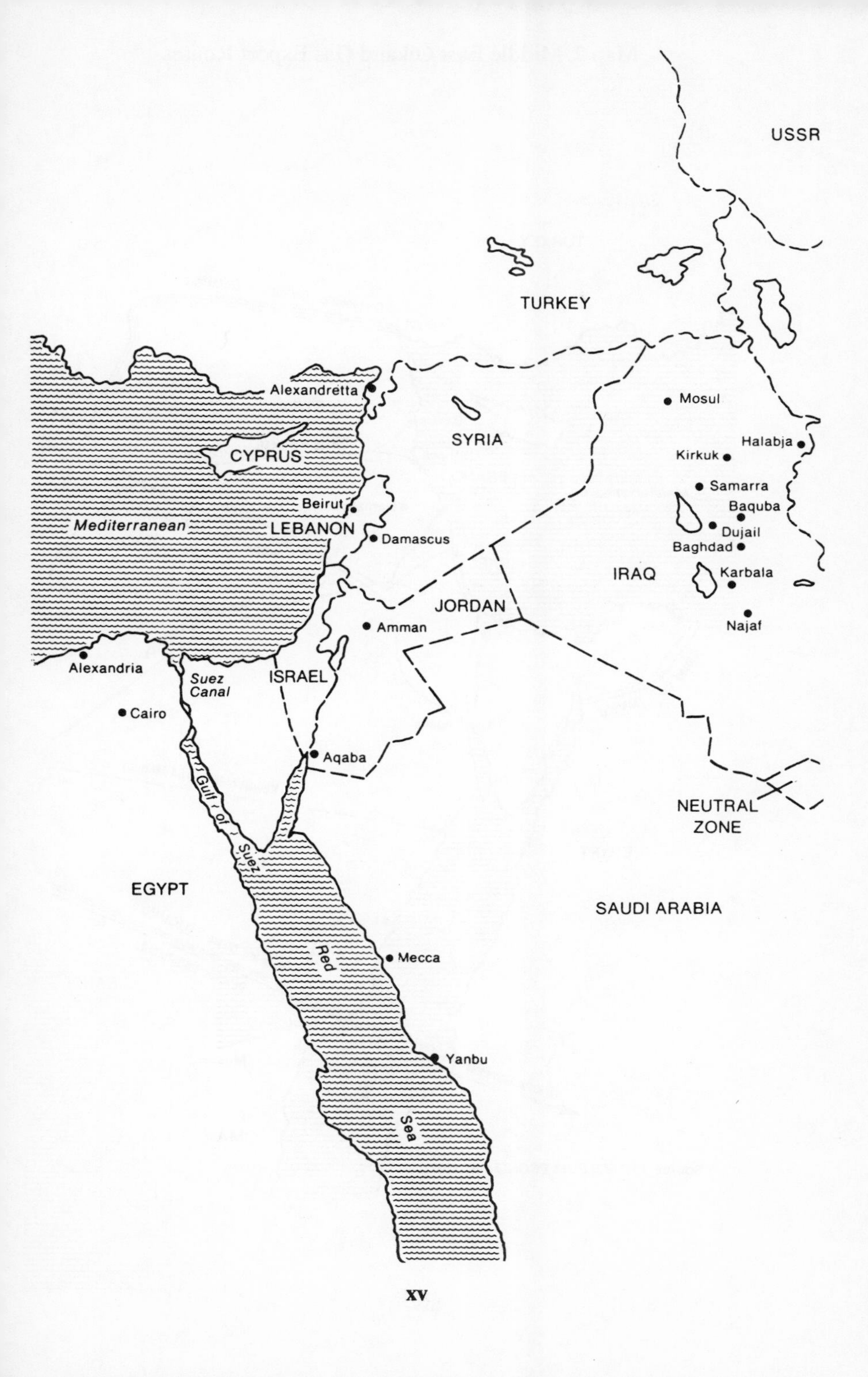
USSR
TURKEY
Alexandretta
Mosul
SYRIA
Halabja
CYPRUS
Kirkuk
Samarra
Beirut
Baquba
Mediterranean
LEBANON
Dujail
Damascus
Baghdad
IRAQ
Karbala
JORDAN
Amman
Najaf
Alexandria
Suez
Canal
ISRAEL
Cairo
Aqaba
Gulf of Suez
NEUTRAL
ZONE
EGYPT
SAUDI ARABIA
Red
Mecca
Yanbu
Sea

Map 2. Middle East Oil and Gas Export Routes

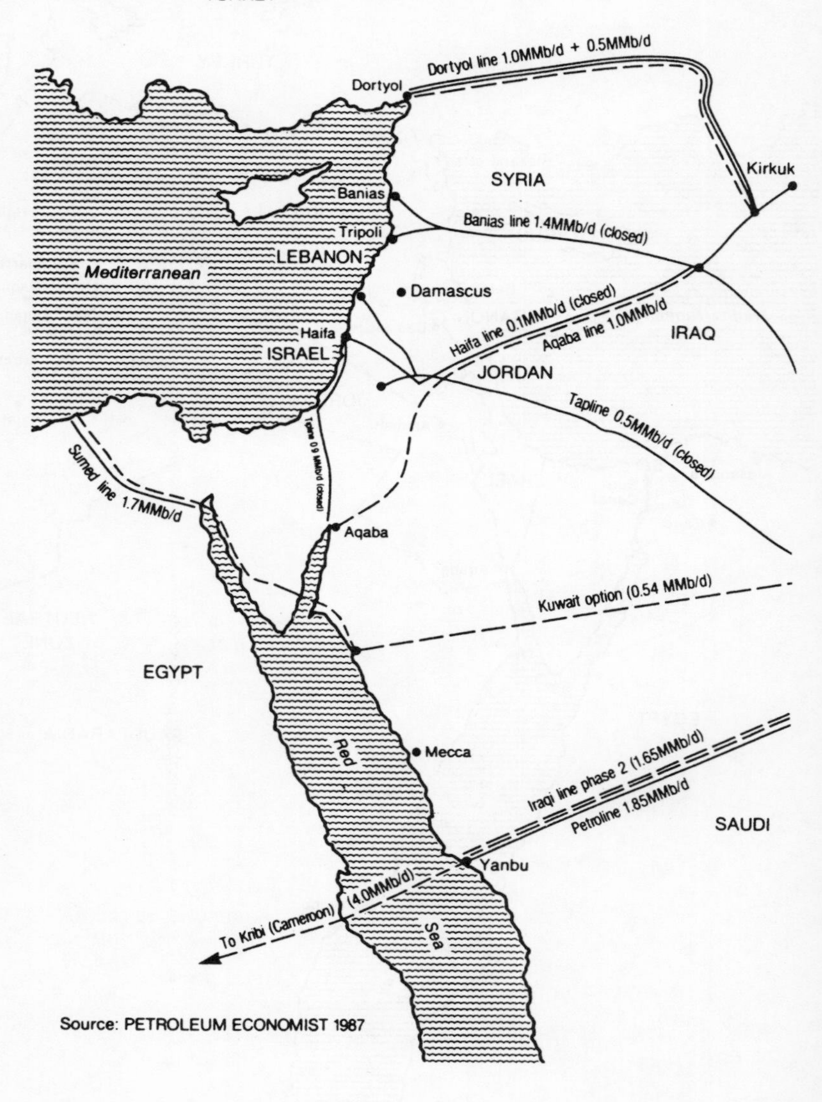

Source: PETROLEUM ECONOMIST 1987

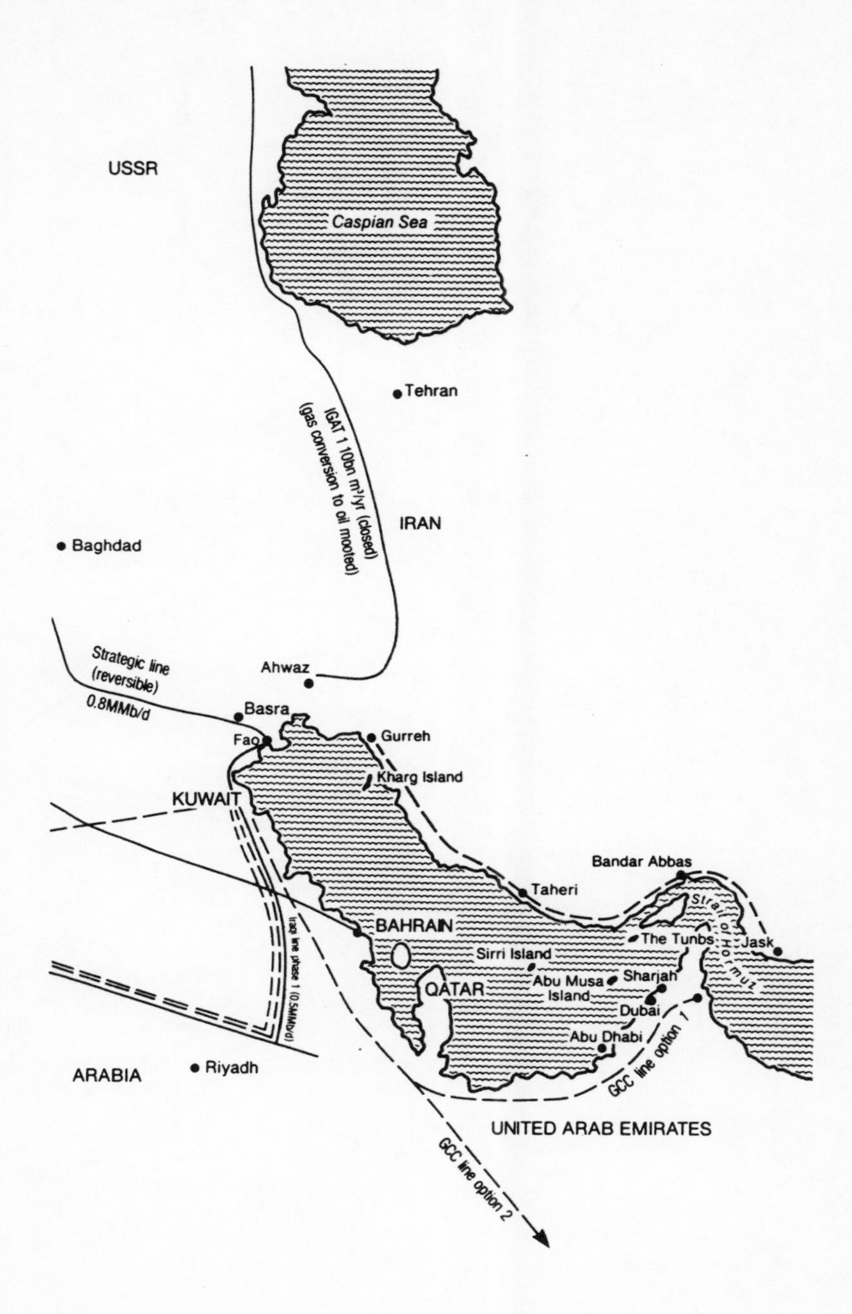
USSR
Caspian Sea
Tehran
IGAT 1 10bn m³/yr (closed)
(gas conversion to oil mooted)
IRAN
Baghdad
Strategic line
(reversible)
0.8MMb/d
Ahwaz
Basra
Fao
Gurreh
Kharg Island
KUWAIT
Iraqi line phase 1 (0.5MMb/d)
Bandar Abbas
Taheri
Strait of Hormuz
BAHRAIN
The Tunbs
Jask
Sirri Island
Abu Musa
Island
Sharjah
QATAR
Dubai
Abu Dhabi
GCC line option 1
ARABIA
Riyadh
UNITED ARAB EMIRATES
GCC line option 2

1. IRAN

Introduction

More than those of any other state in the Gulf, the actions of Iran have a significant influence on the foreign policy agendas of the other littoral states. This tended to be the case during the reign of the Shah. It has been more clearly so since the Iranian revolution more than 10 years ago. Even the foreign policy of Iraq, the other major state in the region, has tended to be mainly reactive to the actions and perceived stances of its large neighbour to the east. Two examples stand out. In the mid-1970s, Baghdad signed the Algiers Agreement as a result of the extensive help Iran had given to the Kurds inside northern Iraq. Some five years later, Iraqi forces rolled across the border in response to the Islamic revolution in Tehran: either to prevent Iranian expansionism or to take advantage of the revolutionary turmoil, depending on the view one takes of the motives of Baghdad.

Whatever the regime prevailing in Tehran, as long as the 'nation-state' of Iran continues to exist it will go on exerting a deep influence on the politics of the area - not least for geographical and demographic reasons. The country's location stretching as it does along the entire 900 mile eastern coast of the Gulf, gives it a direct interest in the waterway from the Shatt al-Arab flowing into it in the north to the Strait of Hormuz at the southern entrance. Moreover, its reliance on the Gulf for the export of its crude oil more sharply focuses that interest. Quite apart from regional political aspirations, this geographical situation compels Iran to pay attention to the actions of the other littoral states which have at least the potential to affect its interests.

Demographically also, Iran dominates the area. According to the 1986 census, the Iranian population was well over 48 million. It not merely has the largest population of all the states on the Gulf, it also has 50 per cent more inhabitants than the seven Arab littoral states combined[1]. This trend in population distribution would appear set to continue, given the youthful composition of the Iranian population[2]. Consequently, the demographic imbalance promises to become more acute; a seminar held in Mashad in September 1988 estimated that by the year 2,022 the population of Iran will top the 140 million mark[3].

These geographic and demographic factors also compel the attention of the two superpowers. Iran shares a 1,300 mile land and sea border with the Soviet Union. Historically, this has made the two states extremely wary of each other. Moscow continues to be anxious about the potential influence which Iran could wield over its co-religionists in the southern republics of the USSR. Iran's fears of the Soviet Union are based on a long history of Russian and then Soviet aspiration to expand southwards and also on Moscow's potential encouragement of national sentiments among its Kurdish and Azerbaijani minorities in the north. The relationship is further complicated by the 520-mile border which Iran shares with Afghanistan, where Tehran supports Shi'a and some Sunni guerrilla movements.

Iran's geographical location gives it added value for the US. Under the Shah, Iran was seen as a bulwark against the spread of Soviet influence in the region. Since the revolution, the US has not been able to rely on Iran as its tool to confront communism. Periodic attempts by Moscow to court revolutionary Iran, whether between 1979 and 1981 or in 1987, have made Washington uneasy. However, it is no doubt comforted by the enduring potential for suspicion and conflict which exists between Iran and the Soviet Union, born out of geography and the religious question. Iran's location is of further importance to the US. For it backs on to arguably the most reliable ally of the US in southern Asia, Pakistan, to the east, while to the west its shadow falls over the moderate Arab Gulf states.

Iran's demographic dynamism is also of interest to the US and the West in general because it offers the possibility of a large and growing market for their goods. Indeed, of the Gulf states with high proven levels of oil reserves, Iran is the only one which also has a large population. Between 1976 and 1980, US exports to Iran were expected to amount to $24 billion, of which up to half were to be non-military goods[4]. Of course the arms purchases of the Islamic regime in the post-war context will be different from those of the Shah's time, while the oil price remains depressed when compared with that of the late 1970s. Nevertheless, when harnessed to Iran's oil export potential, the effective demand could soon act as a fillip to the export industries of the West.

The enduring centrality of Iran, both to the other states of the region and to the superpowers, makes it particularly important to understand the guiding principles, and in particular the foreign policy, of the regime. The problem is that the outside world, and especially the West, is extremely poorly placed to undertake such an exercise. This is only partly explained by the fact that the post-revolutionary regime in Tehran has shown itself antipathetic towards the governments and visitors of many Western states. Indeed, it is uncertain to what extent the West has ever understood the politics of Iran, even during the period when a close relationship existed first with Britain up to the Mussadegh era and then with the United States. James Bill has shown how the upper echelons of the US political and administrative establishment, especially in the National Security Council and the Department of State, were 'consistently pro-Pahlavi partisans'[5]. Sir Anthony Parsons, the former British ambassador to Tehran, is well known for his honesty in admitting his misjudgement of the pre-revolutionary disturbances[6].

The myopia which existed in the diplomatic and political communities was also prevalent in the journalistic field. Dorman and Farhang, in a study of the US press, have taken as their central hypothesis that: 'The major shortcoming of American press coverage of Iran for twenty-five years was to ignore the *politics* of the country'[7]. As well as consistently failing to discover the currents beneath the apparently calm surface, the US press also tended blandly to take at face value the information and interpretations of the US diplomatic and political communities. Their book is tellingly sub-titled 'Journalism of Deference'.

It therefore appears that Iran remained little understood, even during the period when Westerners had relatively free access to Iranian society. In recent years, with the lack of direct diplomatic relations and the unfamiliarity of both the figures and practices of the new regime, this state of affairs has proved difficult to rectify.

Iran and the End of the War

On 16 July 1988 a meeting took place in Tehran of nearly all the central figures in the Islamic regime. They included the Speaker of the *Majlis* (the lower house of parliament), Ali Akbar Hashimi-Rafsanjani, the President, Ali Khamene'i, the Prime Minister, Mir Husain Musavi, Ahmad Khomeini, Ayatollah Khomeini's son, and the available members of both the Council of Experts and the Council of Guardians of the Revolution. They met, on the instructions of Ayatollah Khomeini[8], formally to endorse the decision to accept UN Security Council Resolution 598 and to accept a ceasefire in the Iran-Iraq war.

Rafsanjani told a press conference in Tehran that the decision was not a sign of weakness. Yet, in spite of his protestations to the contrary, that is exactly what the acceptance of the ceasefire was, despite its being a brave political action. The decision thus to end the war represented a *volte face* on the part of the revolutionary regime in an area of policy which had hitherto appeared most sacred.

Since July 1982, when Iraq was driven back from large areas of Iranian soil, the Islamic regime alone had been responsible for the continuation of hostilities. Moreover, it had not simply pursued the war as a means to attain the end of 'punishing the aggressor'. Rather, it had turned the war into an article of revolutionary faith. With the Islamic regime high on rhetoric but vague on substance, the war was the one tangible and inviolable policy of the revolution. Indeed, only days before the decision to end the war, the *Majlis* had passed a bill providing for war against Iraq 'until final victory'[9].

To overturn what had become the core issue of the revolution was an unexpected and shocking event. So grave was this development that most of the leading figures of the revolution had been assembled to endorse the policy in order to prevent the decision-making elite from breaking ranks over it. However, even that was not felt to be enough. In order to give real legitimacy to the about-turn the ailing Imam, Khomeini, gave it his public seal of approval, in a long message read out over Iranian radio on 20 July 1988[10].

The decision to end the war in this way was not made from choice. The words of Khomeini illustrate this very clearly. In his 20 July speech he referred to the acceptance of SCR 598 as a 'very bitter and tragic issue for everyone and particularly me', and spoke of having 'drunk the poisonous chalice of accepting the resolution'[11].

The regime's extreme reluctance to agree to make peace, and then the bad feeling with which SCR 598 was accepted, led some to see it as a purely tactical manoeuvre. The Iraqi information minister, Latif Nusayif Jassim, adhered to this view. He believed the new Iranian posture 'did not proceed from a genuine desire to establish a real, just peace ... and to conclude a permanent, comprehensive peace accord'[12]. While such scepticism was not unreasonable given the background to the conflict, the suggestion that the Iranian move was purely a tactic with which to catch Iraq out proved to be mistaken.

Although the Iranian leadership may not have found the ending of the war palatable, it was a strategic response to the wishes of large sections of the Iranian people. It was the reluctance of the population to continue the conflict that forced the hand of the revolutionary elite. Some evidence of war weariness had existed before 1987, but it was the costly assaults on Basra in the first three months of that year, when between 15,000 and 30,000 Iranians were killed[13], that satiated the mass appetite for the conflict.

The Iranian people had shown their dissatisfaction with the conflict in a number of ways. The most powerful statement of the growing disillusionment with the war had been the popular reluctance to join the various branches of the military engaged in its prosecution. The authorities tried increasingly desperate moves to boost the numbers enlisting. More recruitment centres were established in January 1988. The call-up of students became mandatory for the first time. The President even appealed for more recruits from among the religious leaders (the *ulama*) to act as an example[14]. By the beginning of 1988, when the offensive should have been in full swing, there were less than 250,000 troops opposite Basra, compared with up to half a million the year before[15]. The difficulty in recruitment was effectively admitted by the authorities when they extended the period of conscription in January 1988 from 24 to 28 months to stem the drain on numbers[16]. They were left with no alternative but to make extravagantly false claims as to the level of recruitment[17], and ultimately to present the lower numbers as a change in military strategy[18].

Secondly, even among those who did join up the willingness to fight had all but evaporated. This loss of fighting spirit was most graphically displayed in the spring of 1988 when the Iranian military machine seemed to crumble. Not only were the volunteers incapable of mounting any full-scale attacks, but they had even lost the stomach to defend the positions they held. In rapid succession, the Iranians were driven out of the wedges of Iraqi land which they had previously fought so hard to take. The Fao Peninsula fell on 17 April, to be followed by victory on the Shalamcheh sector opposite Basra on 25 May. The important oil centre of Majnoon Island followed on 25 June. Meanwhile, the National Liberation Army (NLA) sponsored by the opposition group, the Mujahedin-e Khalq, had become increasingly daring in its forays from its bases in Iraq and had briefly held Iranian border towns like Mehran. The apparent collapse of the Islamic regime's military ability exploded the revolutionary rhetoric about punishing the Iraqi President for starting the war. It also revealed that the willingness of the population to fight and die for the regime could no longer be taken for granted.

Thirdly, there was a marked change in the attitude of the population of Tehran, and to a lesser extent those in other cities like Isfahan and Qom, as a result of the new round in the war of the cities between February and April 1988. During this period, the Iraqis unleashed their new 650-kilometre-range missiles. This meant that for the first time Iraq had the capability of striking at will at the urban heartlands of Iran. Altogether between 160 and 200[19] Iraqi missiles were fired resulting in mass civilian casualties. The exact number of killed and wounded is not known, though one British newspaper estimated the number of dead at 1,700 with 8,200 injured[20]. Although these estimates include the casualties on both sides, Iran bore the brunt of the attacks.

Iraq's use of long-range missiles was a further reason for the undermining of morale in Iran's urban centres. The revolutionary government showed an obvious inability to respond both in kind and in proportion to these attacks. This amounted to an effective failing on its part to deter the Iraqis from further attacks. Only some 50 missiles were launched by Iran at Baghdad and other Iraqi cities[21]. Furthermore, most of those falling in the capital landed relatively harmlessly in the less densely populated south-east of the city. The desperation with which many Iranians, especially members of the middle class, left Tehran, was proof of the terror that seized urban dwellers in the Islamic republic. As many as 50 per cent of the inhabitants of the capital were reported to have left for fear of the missile attacks[22].

Finally, Iraq's use of chemical weapons against the Kurdish town of Halabja came as a resounding warning to the Iranians of the means to which the Iraqi regime would be willing to resort in pursuance of its own self-preservation. During the attack on the town in mid-March, some 5,000[23] civilians were reported killed as a direct result of chemical weapons. Their extensive use by Iraq, combined with the introduction of long-range missiles, raised the spectre of such weapons being used in combination against population centres in Iran. The effect of Iraqi use of chemical weapons therefore became comparable to the US use of nuclear weapons against Hiroshima and Nagasaki at the end of the Second World War and, as in the Japanese case, the use of such methods helped to bring an end to the war.

Clearly, then, there is a genuine popular yearning for peace in Iran, particularly in the cities. Whatever the ideological desires of the ruling regime in Tehran, the desire of the majority of the people for peace has prevailed. The importance of the role played in this by Rafsanjani is not that he was in favour of ending the war. As Speaker of the *Majlis* since July 1980, and an important member of the regime since the outbreak of the revolution, he was patently a supporter of the continuation of hostilities in earlier phases of the conflict. Rather, he became convinced of the wisdom of ending the war because he could see its underlying unpopularity, which resulted in a parlous deterioration in Iran's military position at the front. The twin prospects of military collapse and popular dissatisfaction ultimately threatened the continuation of the regime and his position within it.

The idea of the ceasefire as nothing more than an Iranian tactic to obtain a short-term breathing space pending a resumption of hostilities from a position of strength does not stand up to analysis. Given the mood of exhaustion in Iran, the situation would have to change dramatically for the people wholeheartedly to support the resumption of the warfare. It cannot even be assumed that a formal peace which heaped national humiliation upon Iran, such as the restoration of Iraqi sovereignty over the whole of the Shatt al-Arab, would result in reopening the fighting. However, though the short-term effect of such a development might not be to destabilise the region, the longer-term implications of a humiliating settlement could be catastrophic. The enduring sense of injustice that would be an intrinsic part of such a peace could be almost relied upon to lead to another outbreak of hostilities, though probably not before the next century. The Treaty of Versailles may provide a pertinent historical analogy. Germany capitulated from exhaustion, even though little fighting took place on German soil. As the defeated party, the German people were deeply humiliated at the 1919 peace conference and thereafter. It was only some 17 years before a rejuvenated Germany began unilaterally to reverse the most contentious aspects of the peace with the reoccupation of the Rhineland.

In the short term, the Iranian regime appears to have coped well domestically following its sudden decision to accept the ceasefire. The decision to extend the recess of the *Majlis* appeared to be based on the need to minimise criticism of the government and its policies at the time. However, the return of the *Majlis*, following a 39-day hiatus, suggested that the regime was more self-confident. One reason for the absence of any outburst may have been that, in the immediate aftermath of the ceasefire, the Iranian people appeared to be just as surprised and stunned by the development as outside observers. This should not be taken to mean that the regime has not had to pay a price for its *volte face.* Some blame for keeping the war going needlessly for so long is bound to attach itself to the regime, even if this manifests itself more in the form of cynicism than of organised opposition. In particular, the regime has to live with charges that it missed two crucial opportunities to secure a peace on much more favourable terms than appear available in the autumn of 1988.

First, Iran had the opportunity to make peace in July in the aftermath of the Iraqi withdrawal from most of the Iranian territory it had captured earlier. To have done so would have left Tehran in a commanding diplomatic position. Iraq, desperate to see the war end, was in no position to make demands - certainly not, as it has appeared to be doing since August 1988, to seek the return of its sovereignty over the whole of the Shatt al-Arab. An end to the conflict then would also have left Iran as the dominant power in the Gulf, a situation which the smaller Arab littoral states would have had to acknowledge. Moreover, the superpowers would have been obliged to deal with Iran as the pre-eminent regional power, which would have given Tehran considerable latitude to play Moscow and Washington off against each other.

Most important perhaps would have been the sense of credibility and self-confidence which would have been engendered amongst the Iranians themselves. Certainly, the peace would have been achieved before the debilitating human-wave attacks which Iran was to pursue as a military strategy over the next few years. Ultimately, Iran would have been in a stronger position to export its religious and political messages to the Islamic people and the Third World. It would have been more readily perceived as an example of religious purity and political independence. By choosing to continue the war on Iraqi territory, Iran merely descended into the role of a regional threat, and rendered its characterisation as a fanatical monster state all the more convincing.

The second missed opportunity for peace was in July 1987 in the wake of the adoption of Security Council Resolution 598. Iran's general position was weaker than in 1982 because the potency of its armed forces had been called into question. In addition, with the presence of a large foreign naval fleet in the Gulf, essentially aimed at limiting Iranian military options at sea, it was isolated and obviously fighting more than just the Iraqi armed forces. Nevertheless, it still had the military initiative. The previous winter Basra had almost fallen and there was every prospect of another offensive, of similar ferocity, later in the year. Furthermore, Iran continued to occupy strips of Iraqi territory, most notably at Fao. This gave Tehran an additional bargaining counter. The Iranian Government chose not to accept SCR 598, instead procrastinating at the diplomatic level in order to buy time. But the time it bought was only sufficient for its military position to deteriorate profoundly. The measure of the diplomatic blunder is revealed by a comparison of the military circumstances which existed when the Iranian authorities were obliged to accept SCR 598 with those of a year before.

Domestic Politics

When writing about revolutionary Iran, many authors and journalists choose to accentuate the unstable nature of its politics and of the political process. Iran is often presented as a state riven by personal, factional and ideological divisions. The enduring impression which is left by such commentaries is that the Islamic regime is on the point of breaking up, of exploding into a bitter and bloody internal struggle, of collapsing from within.

It is of course true that there are personal rivalries, that coalitions are made and broken and that there are competing policy visions. However, much of the concern about the future nature of the regime has been generated because of Khomeini's age. When the Iranian revolution occurred in early 1979 Khomeini was 77 years old. In 1988 he is 86. While one can be certain that he will not live forever, by continually focusing on the precariousness of his health, many commentators have failed to notice the stability among the political elite.

One of the most striking characteristics of the history of the Iranian revolution is the relative shortness of the period of chronic upheaval. Even though the revolution itself was carried out by a broad coalition of anti-Shah forces, some secular, some Islamic, its religious character was quickly established. Indeed, the concept of *Vilayat-e Faqih*[24], which formalised political authority in the hands of Khomeini, was adopted in the constitutional referendum of December 1979. Out-manoeuvred at the political level, many of the opposition groups turned increasingly to the use of force. Despite considerable success in this strategy, which included the assassination of many leading figures of the revolution in August 1981, the use of violence did not undermine the regime. By the end of 1982, the wave of violent attacks had ceased. In February of the following year, the regime demonstrated its self-confidence by moving against the formerly powerful Tudeh Party, by then the most important opposition group not to have broken with the regime. Even in the immediate aftermath of the acceptance of SCR 598 the streets of the capital remained quiet. The absence of an obtrusive security presence in Tehran illustrates the extent to which the regime is confident that there will be no outbreak of violent opposition.

The continuity and stability of the Islamic regime are perhaps best illustrated by the longevity of its leaders. Of the leading figures of the revolution, most have held high office for considerable periods of time. Rafsanjani emerged as the *Majlis* Speaker in July 1980, and was re-elected for the third time in 1988. Khamene'i was elected President in October 1981 and was returned for a second term in 1985. Musavi was chosen as prime minister in late October 1981. There also appears to be a marked degree of continuity amongst the technocratic elite of the revolution.

Much has been made of the political opportunism of individuals in the Iranian leadership. However, it would be misleading to portray these men as the carpet-baggers of the revolution. Each one had proved himself a dedicated opponent of the Shah and supporter of Khomeini well before the revolution. Many of the politically prominent *ulama*, such as Rafsanjani, Khamene'i and Ayatollah Montazeri, Khomeini's formally acknowledged successor, had studied under Khomeini at Qom. Moreover, many of these figures had personally suffered before the revolution as a result of arrest and harassment by Savak, the Shah's secret police. For instance in March 1975, on the eve of the formation of the Shah's new mass mobilisational party, the Rastakhiz, Khamene'i and Montazeri were both arrested[25]. In other words the senior figures of the regime do possess a marked degree of personal legitimacy, both because of the roles which they played in helping to topple the Shah and also in their close attachment to Khomeini.

The backdrop to the present developments in Iran is therefore one of relative stability and continuity. This perhaps goes a long way towards explaining the surprisingly little fall-out from the momentous decision to end the war and negotiate directly with senior members of the hated Ba'thist regime in Baghdad. Certainly the aftermath of the acceptance of SCR 598 revealed the weakness of opposition forces outside the regime. It also displayed the lack of an organised alternative to the incumbent leadership.

When the authorities accepted the ceasefire, it was an admission of the failure of their war strategy over a number of years. It also left the regime momentarily subdued and therefore vulnerable. The opposition group best organised to exploit this situation was the Mujahedin-e Khalq, led by Massud Rajavi, now in exile in Iraq. Having already experienced some success in skirmishes with the forces of the regime in the Iraqi border area, Rajavi decided 'to go to Tehran'. In doing so he gambled both on the collapse of the regime and on the fact that the Iranian people, tired of war and disillusioned with the economic and cultural austerity of the government, would respond to the military push. The military wing of the

Mujahedin, the NLA, advanced almost to the regional centre of Kermanshah, some 90 miles into the interior of Iran.

Hindsight shows that Rajavi was over-sanguine in his assessment both of the potency of his own appeal.and of the weakness of the regime. Dealing with the first point, his decision to move the Mujahedin's headquarters to Iraq, following his expulsion from France in the summer 1986, was clearly a two-edged weapon. While it gave the NLA the opportunity to organise itself as a fully fledged army and to launch raids directly against the forces of the Islamic regime, it branded the organisation with the stigma of collaborating with Iran's enemy. Despite Rajavi's protestations of political independence from Iraq, there can be no doubt that this link has made the Mujahedin considerably less attractive to the Iranian people. As a result, there was not the spontaneous outburst of support for the NLA move that Rajavi might have hoped for.

Moreover, the response of the regime was perhaps more vigorous and coherent than might have been expected, although the manner in which the authorities rallied their military forces did appear to be tinged with desperation. Universities were closed and sporting fixtures postponed as the regime attempted to mobilise a force capable of stopping the NLA. A combination of the weight of numbers and the fact that the NLA's lines of communication were hopelessly over-extended appear to have been crucial in the defeat of the Mujahedin. Without the air support vital to give their troops the necessary cover, the NLA force had no choice but to retreat after sustaining very considerable losses[26].

The regime skilfully exploited this military victory to its own advantage. First, it provided a warning to Iraq that the Iranian armed forces had not completely collapsed, and that the regime would not fall apart. Clearly these were important factors in Baghdad's decision to agree to a formal ceasefire, having procrastinated until then. Secondly, it gave the Islamic regime a timely boost during a period of acute demoralisation. Thirdly, it also acted as a warning to other opposition elements both inside and outside Iran as to the vitality of the regime. In this respect it seems to have been successful as a deterrent to action from other groupings. Fourthly, it gave the regime the excuse to act against opposition elements inside the country, especially those from the Mujahedin. The most obvious manifestation of this was a new wave of executions of political opponents in the autumn and winter of 1988.

The weight of the blow dealt to the NLA thus appears to have boosted the self-confidence of the regime and strengthened it *vis-à-vis* its opponents. Even by their own admission, the Mujahedin lost 1,000 soldiers[27]. Other reports have estimated the number killed on the NLA side as being well over 4,000[28]. Whatever the exact numbers, these are heavy losses for a force which is believed to have been only some 6,000 to 8,000 strong[29].

There now no longer appears to be any organised opposition capable of seizing power or precipitating the downfall of the regime. Of the other opposition factions, the various monarchist groups are small and have no popular base of support; the leftist groups are hopelessly fractured and their political message is obscure; and the Tudeh Party, though having the benefit of a radio station which operates from the Soviet Union, has been gutted by the arrest and exiling of its leading activists. In the absence of any capacity for mass mobilisation, the only strategy for all opposition groups appears to be to wait in the hope that the regime in Tehran will weaken further. Some might be tempted to resort to political assassination, but this would be a sign of desperation.

The absence of organised opposition or a realistic alternative to the regime should not be mistaken for popular support, however. The war has steadily eroded the general backing for the regime over a number of years, and its exposure as a pointless failure will probably deepen the gulf which has opened up between the regime and the people. The trend, then, appears to be back towards the traditional Shi'a notion of government which, in the continued absence of the hidden Imam, leads to all government being regarded as fundamentally illegitimate, even government by *mullahs.* Rather than being a prescription for further revolution, such an analysis has in the past resulted in precisely the opposite effect, as the *ulama* have taken the view that 'bad government is better than no government'[30]. In practical political terms, this fatalistic quietism amounts to legitimacy by default for the present regime. It is not the same as genuine support for the continuation of the regime in power.

This means that, for the short and most probably the medium term, regardless of the personnel of the regime, the broad ideology, symbols and values of the Islamic revolution will continue to prevail. Iran is therefore likely to continue to be an Islamic republic at least in name, and the form of the regime will continue to be a religious one. Most importantly, the political elite is likely to continue to be drawn predominantly from the *mullahs.* If there are to be political upheavals within Iran, then they are far more likely to come about as a result of friction within the regime.

In the past, a broad consensus existed amongst the revolutionary political elite as to the desirability of the continuation of the war. This consensus and the domination of the war over the internal political process meant that difficult choices about the domestic road the revolution was to take could be avoided. As Shahram Chubin wrote prior to the ceasefire, 'the war is not so much the cause of immobilism but rather a substitute for domestic politics'[31]. With the war apparently at an end, the revolution is once again left with few achievements other than the overthrow of the Shah. The two major though related questions of substance now facing the regime on the internal front are socio-economic reform and the process of reconstruction. While there is a general feeling of expectation that the regime finally has the opportunity to shape the country according to a revolutionary prescription, the enduring impression is that there is no coherent blueprint for the future.

The issue of the organisation of the economy and the interests towards which resources are allocated has been one of the underlying questions facing the revolution from the onset. As Shaul Bakhash put it: 'The revolutionaries in Iran came to power with ideological baggage that implied an economic as well as a political transformation of society'[32]. The key questions to have exercised the regime within this broad theme are: the extent to which rural land reform takes place; the profile which the state plays in the productive and distributive economy; and the degree to which the merchants and traders of the bazaar are allowed to operate unfettered.

The early years of the revolution saw a rapid increase in the government's interference in the economy, during which 'the economic role of the state was greatly swollen and that of the private sector greatly diminished'[33]. However, very little of the economic reform legislation was implemented because of pressure from the bazaar[34] and the obstruction of the conservative *ulama*. In the case of the latter, the constitutional structure of the legislature, with the Council of Guardians charged with ensuring that new bills endorsed by the lower house do not violate Islamic precepts, maintained a virtual block on all radical legislation. One such example was the defeat of the bill on the nationalisation of foreign trade in early 1982.

The growing hardships of the war encountered in 1987 seemed to imply that the regime would have to respond with populist economic reform in order to maintain its standing. In particular, inflation had been running at around 30 per cent in 1986[35] and was as high as 50 per cent in 1987[36]. In response to, and exacerbating, the rate of inflation was the practice of chronic hoarding, especially by those merchants and speculators keen to maximise profits. With dissatisfaction growing with this state of affairs, it seemed that the revolutionary rhetoric in favour of the oppressed might be turned into substantive policies. A new body, the Assembly for Determining Governmental Laws, was widely believed to have been created to circumvent the legislative impasse. It was expected to act on such issues as tax reform, commodity distribution, the ownership of agricultural lands and labour legislation[37], all of which had been blocked by the Council of Guardians. Such reforms might have been expected to be more generally popular, and to appeal to the more egalitarian members of the regime. What is clear is that they would have been likely further to antagonise the conservative religious establishment and the bazaaris.

The declaration of the ceasefire seems to have dispelled these hopes for radical, *etatist* reform. With regard to the reconstruction process, the radicals now appear to be firmly on the defensive and a cautious policy of moderation has emerged. The ceasefire has also quietened the clamour for reform to mitigate the hardships of the bulk of the population. The end of hostilities has eased the pressure on the Iranian currency, while opportunities for speculation have decreased. For these reasons alone, a reduction in the rate of inflation would appear probable.

However, the apparent return to a strategy of more emphasis on private enterprise and private property seems to be prompted less by the sudden improvement in the economic environment. Rather it seems to reflect a more active desire on the part of the regime to widen the base of its support amongst its natural constituencies. These include the traditionalist *ulama*[38], who support a free market and resist the notion of increased state intervention in the economy, partly out of Islamic ideological conviction and partly out of economic self-interest, for the 'high ranking ulama are amongst the wealthiest classes in Iran'[39]. The other natural allies of the regime are the bazaaris. Indeed, because of the great respect with which merchants are held in Muslim society, many *ulama* are also wholesale merchants. The diversion of the regime away from the pursuit of *etatist* policies heralds a strengthening of this alliance.

If the radicals in the regime seem to be losing out in relation to domestic economic organisation, the same appears to be the case *vis-à-vis* the reconstruction process. One of the central reasons for Premier Musavi's attempted resignation in the late summer of 1988 was that the regime seemed to be set on a course of accepting help from the West in the reconstruction process, a course which he did not favour but was powerless to affect. Musavi had advocated a more autarkic model for economic recovery, based on the utilisation of skills developed as a matter of necessity owing to Iran's isolation during the war. This brought him into conflict with Rafsanjani, who criticised those who would 'keep the people in a state of abstinence and shortages so that the people are even short of basic goods'[40].

The position of the radicals in relation to both domestic economic policy and reconstruction appears to be one of weakness. Indeed, in the next section, it will be argued that the same is true with regard to international relations. Of crucial importance here is whether this is a temporary reversal or whether the days of the radical nature of the regime are over. The continued existence of well known radical figures[41] in senior positions means that radical policies cannot be ruled out from making a comeback. This might be the case if Khomeini, the one man capable of making authoritative decisions, were to die soon. However, it also applies as long as Khomeini continues in power. This conclusion has much to do with Khomeini's style of leadership, as well as personal and ideological factors.

During the course of the revolution Khomeini has not discouraged factionalism in order to retain the need for a mediating authority, which is where he is able to wield most influence. On the personal side, Khomeini enjoys a very warm relationship with some of the radical leaders. A good example is that of the Minister of the Interior, Ali Akbar Mohtashemi. He was a member of Khomeini's household in Najaf before the revolution and was appointed to the influential home affairs portfolio on Khomeini's instructions. In ideological terms, there are many indications that Khomeini is uncomfortable with the present direction of the revolution. He was intensely distressed by the need to end the war, and continues to espouse a policy of bitter hatred of Saddam Husain of Iraq. Yet, in spite of all this he has endorsed a strategy of moderation in economic and foreign affairs since July 1988.

Despite the continued presence of radicals inside the regime, it seems most likely that the newly adopted moderate policies will at least be given a chance of bearing fruit. Evidence for this can be seen in the new coherence in policies emanating from Tehran. Rafsanjani appears, at least temporarily, to have established a firm grip on decision-making. He seems to have added to his considerable abilities as a power broker by forming an alliance with Khomeini's son, Ahmad, to ensure favoured access to the Imam. His appointment in early June 1988 to the post of commander-in-chief of the armed forces, with a mandate to reorganise the military, has given him considerable power and patronage. Even the newly elected third *Majlis*, which was supposed to contain more radical deputies than ever before, appears to be not ill-disposed towards the Speaker. Indeed, in the *Majlis* vote of confidence in the individual cabinet members, even someone with impeccable radical credentials like Mohtashemi, perhaps Rafsanjani's most implacable opponent, came close to being defeated.

The apparent consolidation of Rafsanjani's power has been more than matched by a decline amongst the radicals. In retrospect, this may be said to have begun with the arrest in 1986 and execution the following year of Mehdi Hashemi - a prominent radical attached to Ayatollah Montazeri's office who was in charge of the Centre for Exporting the Islamic Revolution. His downfall came about as a result of his attempts to embarrass Rafsanjani for his prominent role in the ill-fated 'arms for hostages' initiative pursued by a small but senior clique from the administration in Washington. Since then, other prominent radicals, such as Mohtashemi, have been blighted by their identification with disastrous incidents abroad, such as the Gordji affair in France. The probable complicity of certain radicals in the highjacking of the Kuwaiti aircraft in April 1988 also damaged the standing of this faction. This action proved to be deeply counter-productive, not least because it instantaneously diverted international outrage away from the Iraqi regime's use of chemical weapons against the town of Halabja.

Ultimately, the main reason for the decline of the radicals has been the patent failure of the policies which they tend to advocate. The most obvious one has been the war. Exhortations to prolong the war were quite plainly unrealistic once Iran's inability to continue fighting had been exposed. Perhaps the biggest loser from the ignominious way in which the war ended was the Islamic Revolutionary Guards Corps (IRGC). At the start of the revolution the IRGC was an elite organisation, membership of which was highly selective and involved candidates having to prove their ideological credentials. By early 1988, however, conscripts were being allocated to serve in the IRGC[42]. It was the IRGC which was most closely identified with the military defeats just prior to announcement of Iran's acceptance of SCR 598. This ultimately led to the IRGC minister, Muhsin Rafiqdust, failing to obtain a vote of confidence from the *Majlis* and losing his portfolio in September 1988.

Regional Relations

Underlying the decision to accept the ceasefire was a fundamental shift in principle in the country's foreign policy, namely, that national interests had been established as the core motivation and consideration in the calculation of its international actions. The establishment of the undisputed centrality of the national interest came at the expense of the ideology of pan-Islamic revolution. In accepting the principle of making peace with Ba'thist Iraq, much of the revolutionary regime's Islamic rhetoric became inversely applicable. If the war had been fought against Saddam Husain because, as Khomeini accused, he was trying 'to destroy Islam', then peace would permit him to continue doing so inside Iraq. If, again in the words of the Imam, 'To compromise with oppressors is to oppress', then the Iranian regime has now become an oppressor. If the war had been against blasphemy, as Musavi maintained, then the peace permits blasphemy to continue to flourish[43].

An important aspect of this profound change in Iranian foreign policy is the effect it will have on the Shi'a communities in the Gulf states. The potency of the opposition Shi'a groups in Iraq, Kuwait, Bahrain and Saudi Arabia had already been called into question earlier on in the war by their inability to mobilise the communities in which they exist. This was the case even when Iran had the upper hand in the conflict, and in spite of the fact that Tehran has continued to give active support to such opposition groups. Apart from the abortive coup in Bahrain in 1981, some disturbances in the eastern province of Saudi Arabia in the aftermath of the Iranian revolution, and the periodic, but often minor, bomb attacks in Kuwait[44], the extent of the Shi'a challenge to the ruling Sunni regimes in the Gulf states has been limited.

One can assume that Iran's decision meekly to accept the ceasefire will come as a shattering blow to the morale of these Shi'a groups. Moreover, the removal of the possibility of the military defeat of Iraq and the direct threat to Kuwait which would have ensued is likely to make such groups less attractive to potential recruits, who are now less likely to throw in their lot with a movement whose time appears to have passed. Rather than court persecution at the hands of the prevailing governments, the traditional opposition leaders look set to seek to improve their positions within the existing systems by means of co-operation and negotiation.

The further marginalisation of these Shi'a groups in their own communities would in turn be a persuasive argument for Iran to reduce its support for them. Tehran may well decide that it is in its interest to do this in any case. Support for the implacably alienated groups within other states[45] was a sound enough strategy while Iran was hostile to these states. However, the position of the Iranian Government has changed. It now accepts peace and co-existence with Iraq and desires to have conventional relations with the states of the region and further afield. Suddenly, vigorous support for small, ineffective and embittered groups appears not only anachronistic but positively counter-productive.

For example, were Iran to seek to continue the fight against Iraq through the use of Shi'a proxies, it could provoke a swift and severe military backlash from Baghdad. The response would depend upon the groups which Iran continued to support and the means which were used. Were practical assistance, most crucially arms, training and money, to be dispensed rather than merely propaganda, then Iraq might choose to interpret this as active interference in its internal affairs. Baghdad would also be likely to be more sensitive to the support of groups like al-Dawa, which are committed to a path of violent opposition, than if Iran confined its backing to, for example, the relatively benign and ineffective Supreme Assembly of the Islamic Revolution of Iraq (SAIRI).

With Islamic symbols and slogans such an essential part of the ideology of the Tehran regime, its notional encouragement of the Shi'a in neighbouring states will remain. Furthermore, Iran is unlikely to be so capricious or its new flexible leadership so rash as to abandon completely those opposition figures, like Muhammad Baqer al-Hakim, the head of SAIRI, who are beyond reconciliation with the ruling regimes in the area. The cynical repatriation of such men to their incarceration or death would, *inter alia*, deeply offend many Shi'a activists in Iran, as well as fellow travellers around the world. It would also give valuable ammunition to those more radical personalities who seek to change both the policy and the senior personnel of the regime. In *realpolitik* terms, the threat of the reactivation of these groups could prove to be a useful future weapon at the margins of diplomacy. However, Iran's commitment to such men may no longer go much further than offering them sanctuary, moral support and a personal stipend.

If Iranian national interest rather than pan-Islamism are once more to be the ruling factor in the formulation of foreign policy then, as with Iraq, it will be profitable to consider what those national interests are.

<u>To establish peace in the Gulf waterway.</u> The vast majority of Iranian imports and exports, including all its crude oil sales, move through the waterway of the Gulf. This is likely to continue to be the case with all Iran's oil infrastructure and most of its port facilities lying inside the Gulf. The original extension of the war down into the Gulf was perpetrated by Baghdad in order to damage Iran in retaliation

for its prosecution of the land war. However, in blunder after blunder Tehran succeeded in entrenching the war in the Gulf and even ensuring in 1986 that the focus of the conflict shifted to the waterway. Indeed, it was Iran's decision to target Kuwaiti shipping which resulted in the arrival of a high profile US fleet inside the waterway. Iran now wants to reverse these trends. The best way to do this involves demonstrating that the Gulf is no longer an area of instability, and that the security of the waterway is assured. Only then will the foreign fleets be radically scaled down. The *sine qua non* of this must be that the ceasefire holds.

To maximise its oil revenue. In the aftermath of the war, Iran will face the challenges of economic reconstruction and of purchasing modern systems for its armed forces. Both require large amounts of cash, even if it becomes appreciably easier for Iran to obtain international credit. The rebuilding of the economy in particular will be a massive operation. Firstly, in the short term, as Rafsanjani has said, the regime must deal with the people's immediate food and medical needs[46]. Secondly, there are those areas which have been damaged directly by the war. These include urban areas shattered by missiles, bombing and artillery shelling. They also include building up the industrial capacity which, whether oil installations at Kharg Island or power stations in the centre of the country, were the specific targets of Iraqi bombing raids. Thirdly, there are those projects which were slated to be constructed, but which had to be postponed owing to the shortage of foreign exchange. The regime's problem is that it now has several top spending priorities. Moreover, it has portrayed the reconstruction process as a cure-all palliative for the future. A shortfall between the extent of the promised economic regeneration and popular expectations could further weaken the regime's popularity and the fortunes of moderate policy. The size of Iranian oil income, together with financing from other quarters, will be of central importance to the well-being of the regime and its current set of policies.

To ensure that Iraq does not dominate the Gulf region. If Iran's initial defensive campaign against Iraq in 1980 was partly aimed at preventing Baghdad from establishing regional hegemony, then this objective has not changed. Tehran's traditional view of the Gulf is that, for geographical reasons, it is Iran which should play the dominant role in the area. Historical experience, such as the US 'twin pillars' policy, has tended to reinforce this view. It is not only this traditional view of the Gulf which makes Iran smart at the idea of Iraq extending its influence. Iran fears that Iraq will try to bring the Gulf states firmly under its influence, and that this may even include an attempt by Iraq to join the Gulf Co-operation Council. If Baghdad were indeed to attempt such a strategy and to be broadly successful, then it would effectively freeze the majority of the Gulf states in the anti-Iranian posture which prevailed increasingly in the mid 1980s. Iran would fear that Iraq would attempt to harness this Arab Gulf alliance to threaten its national interests.

To establish normal diplomatic relations with both superpowers. It is obviously in Iran's interests to have a sound and stable inter-state relationship with both the Soviet Union and the United States. The primary rationale for this is its position as a major regional power in an area of great interest to both the Soviet authorities because of their proximity, and the Americans because of their preoccupations over oil supplies. Trying to confront either superpower, as Iran found to its cost in the Gulf in the autumn of 1987 and the spring of 1988[47], is a hopeless task. The only way for Iran to maximise its independence of action is by exploiting this regional interest, and playing on great power suspicions. A balanced relationship with the superpowers will also have the advantage of drawing each side into effectively guaranteeing Iranian independence in order to prevent the other from acquiring the sort of unhealthy relationship which typified US-Iranian links under the Shah.

There can be little doubt that Iraq will continue to be perceived as the greatest potential threat to Iran for some time to come. This view will endure as long as the Iraqi armed forces have a massive material advantage. The Iraqi military has built up a large stock of hardware during the course of the war. By contrast, the Shah's massive arsenal of arms and equipment has long since been used up or has gone to waste owing to the inability to buy spare parts. As a result of these two opposite trends, the disparity in the weaponry of the two sides is stark. The position with regard to combat aircraft illustrates the point. In 1987, Iraq had between 400 and 500 modern, well-equipped war planes. The figures for Iran's air force a year earlier were 80 to 105 aircraft[48], many of which were older and inferior to those of Iraq. Iran has a lot of costly ground to make up before its armed forces become sufficiently strong to act as a deterrent against punishment strikes from Baghdad. Given the likely continuing weakness of the oil price and competing areas for expenditure within Iran, Iraq may continue to enjoy a decisive military superiority well into the next decade.

Iran is therefore likely to have to rely increasingly upon diplomacy to protect its vital interests. This imperative will tend to underscore the necessity of mending its fences and seeking to end its diplomatic isolation. Most important in this respect will be relations with the Gulf states, the superpowers, Europe and Japan. However, multilateral international forums could also emerge as stages on which the two rivals seek to weaken each other's position. The United Nations is an obvious example, but the scene of the most intensive manoeuvring could be the Islamic Conference Organisation and the Non-Aligned Movement.

Iran is certainly looking to improve its relations with the GCC states. During the course of the war Tehran's stance towards the Gulf states as a whole fluctuated wildly between threats and appeals. Most of the states themselves, fearful of the unpredictability and zealotry of the regime, gave corporate support to Iraq through the GCC. Even so, in the eyes of the individual GCC members, both combatants were viewed almost equally as enemies. It is thus with a sense of unease that they have watched the Iraqi regime unexpectedly emerge from the conflict brimming with self-confidence. Consequently, at a time when Iran is eager to garner as many friends as it can, the Gulf states are likely to be more responsive to its overtures.

It is, however, easier for Iran to develop good relations with some GCC states than with others. Broadly speaking, there are two groups. The first, consisting of Oman, the United Arab Emirates, and to some extent Qatar, either have historically good relations with Iran, or at least no major difficulties to surmount prior to a strengthening of ties. It is Oman with which Iran has the best relations, which may almost be regarded as cordial. In spite of the fact that both countries share sovereignty over the waters running through the Strait of Hormuz, there were no confrontations between their respective navies, even during the peak of the tension at sea. The Omani Minister of State for Foreign Affairs, Yusif bin Alawi, has a good personal relationship with the Iranian leaders. He visited Tehran for a week just after the ceasefire was announced, and returned in early September when, amongst others, he held talks with Rafsanjani and Musavi[49]. Oman is a particularly good friend for Iran to have at the GCC because of its deep suspicion of Iraq, which dates back to the aid which Iraq gave the Dhofari rebels in the Sultanate in the 1970s, and the practical help rendered by Iran in quashing the revolt.

Iran also enjoys good relations with two of the emirates in the UAE. The links with Dubai, which have been cultivated by its ruler Shaikh Rashid, have principally been founded on strong trading ties, which endured even in the midst of the Gulf conflict. The importance of the Iranian market to Dubai is illustrated by the fact that in 1986 re-exports to Iran were around the $350 mn mark, about double the value of goods re-exported to the next most important destination[50]. The presence in Dubai of about 80,000 people of Iranian origin, amounting to some 20 per cent of the total, deepens the relationship. The links between Sharjah and Iran are also based partly on demography, and partly on shared ownership of the Mubarak oil field. This relationship with the two emirates means that Tehran's position can always be given a hearing in the federation. The important conduit which this provides was recognised by the GCC when it made the UAE the formal interlocutor with Iran after the GCC summit in Riyadh in December 1987.

Qatar may also be placed in the category of having reasonable relations with Iran. During the war, Iran would often send envoys to Doha, and the amir, Shaikh Khalifah, was always willing to be conciliatory towards Tehran. Indeed, Qatar refused to give base facilities to the US navy during the months when they were escorting Kuwaiti reflagged vessels. It even refused to permit part of a navigational system, to be paid for by the Japanese, to be sited on its territory, such was the urgency not to offend Iran. Of course, Qatar continued to endorse anti-Iranian resolutions at the GCC. This organisation gave Doha the anonymity it needed in order not to be seen as directly working against Iranian interests.

An improvement in relations with the second group within the GCC, comprising the upper Gulf states of Kuwait, Saudi Arabia and Bahrain, will be more difficult to engineer. Iran has had substantive differences with each of these three states during the period of the conflict. Any improvement in relations could be viewed with suspicion by those who take a more radical perspective on Iranian foreign policy. Both Kuwait and Saudi Arabia are regarded as the states which kept Iraq from economic collapse. Both gave financial help to Iraq in the early stages of the war, and more recently they have sold the crude oil from the neutral zone on Iraq's behalf.

Iran also has separate grudges against the two states. It was Kuwait that legitimised the active presence of the US fleet in the Gulf through the re-registration of half its tanker fleet under the American flag. From July 1987, the US fleet organised convoys of ships flying the stars and stripes up and down the waterway. The large US naval presence resulted in a number of clashes in which the Iranians fared the worst, and ultimately, in the shooting down of a civilian aircraft on a flight from Bandar Abbas to Dubai.

The main barrier in the way of an improvement in Saudi-Iranian relations relates to the pilgrimage. In July 1987, 402 pilgrims were killed in Mecca, the majority of them Iranians. The exact circumstances leading to the killings remain obscure. What is beyond dispute is that both sides believed themselves to be in the right. The Iranians accused the Saudi authorities of ordering their forces to open fire on the pilgrims. The Saudis say that the Iranians were marching on the Grand Mosque and were apparently intent upon insurrection. Whatever the facts, the incident has made each side extremely wary of the other. Fears that the Iranian regime would try to disrupt the 1988 pilgrimage helped lead to Riyadh severing diplomatic relations in April 1988. The recent deterioration in relations between the two states is in turn underpinned by a substructure of suspicion between the Wahhabi and Shi'a movements within Islam.

Nevertheless, there is no doubt that the will to improve relations currently exists on both sides. After the 1988 pilgrimage, which was boycotted by Iran, King Fahd went out of his way to offer an olive branch by expressing regret at the absence of the Iranians[51]. Subsequently, indirect talks have been opened up using Pakistan as an intermediary. Even when the diplomatic ice is broken, matters of substance may continue to block an improvement in bilateral relations. Foremost amongst these is Iran's future participation in the *haj*. In response to the events the previous year, Riyadh succeeded in getting the backing of a meeting of the Islamic Conference Organisation strictly to limit the numbers of pilgrims coming from different countries. Iran's quota was put at 45,000, well below the 150,000 minimum the Iranian authorities wanted to participate[52]. It was Ayatollah Khomeini himself who announced that, given those terms, Iran would boycott the pilgrimage. It therefore seems that if Tehran is to improve relations with Riyadh it must make further painful concessions over the number of pilgrims it sends on the *haj*.

Iran's most recent quarrel with Bahrain is centred on the extensive practical support which the island state gave the US navy during the latter stages of the conflict. The facilities granted to the US mushroomed in 1987 from a modest Administrative Support Unit to extensive and highly visible sea and air back-up. Though the crisis itself is now over, it is likely that the US will seek to consolidate these infrastructural provisions. A return to the small ASU facility is therefore improbable. In that case, Bahrain will continue to provide the advanced land basing facilities which the US would require if it was to return to the area speedily and in force. The Iranian regime is likely to feel that it is the target of such provisions, and that their existence erodes its standing in the area.

The Superpowers

Iran's new-found desire for acceptance as a full and constructive member of the international system promises a general improvement in its relations with both superpowers. This has already begun to manifest itself. According to the State Department there have been contacts between Iran and the US through third parties since the spring of 1988. There have also been a spate of visitors in Tehran from the Soviet Union.

The superpowers should not misconstrue the attitudes underlying this new posture on the part of Iran. They first of all amount to an admission that the aggressive and confrontationist stance adopted towards the US in particular was not a success. Iran simply conceded the propaganda war to Iraq as far as the world diplomatic community and much of global public opinion was concerned. Moreover, it needlessly drew the Americans into a direct military confrontation which it was unlikely to win. Not only did the regime virtually take on the US, but its actions also established the environment for extensive co-operation between the superpowers aimed specifically at curbing Iranian ambitions. The change in the form and substance of Iranian foreign relations effectively represents a realisation by Tehran that it has more chance of achieving its foreign policy goals through a more constructive approach to international relations than was evidenced in the earlier days of the revolution.

The desire for normal diplomatic relations with the superpowers in no way diminishes the determination that neither should come to dominate Iran's internal affairs. This is a principle to which the regime and, one may assume, the majority of the people subscribe. The country's history has too often been punctuated by occupations, externally sponsored demonstrations and putsches, and improper conduct on the part of foreign embassies for Iranians not to want fiercely to protect their newly acquired independence. During the era of the Shah's rule, national indignities reached a peak with, for instance, US military personnel in Iran subject not to local laws but to their own. It is with this domination and interference fresh in his mind that Ayatollah Khomeini continues to reiterate the slogan of 'neither East nor West'.

Lumping the two superpowers together often tends to suggest that Iran's existing and potential relationship with both of them is equal in nature. This is not the case. As things stand in the winter of 1988 it is clear that the relationship with the US is in more of a trough than that with the Soviet Union. The relationship was more difficult in the early days of the revolution because of the role which Washington had played in supporting the Shah. This was directly worsened by the storming of the US embassy and the holding of its staff as hostage until the beginning of 1981. While the affair may have exorcised some of the bitterness and resentment which had built up in Iran over the past conduct of the US, it created bitterness and resentment towards Iran on the part of the American people. The barrier to improved relations exists as much on a popular level in the US as in Iran. This American ill will towards Iran was reinforced by the national humiliation of the 'arms for hostages' deal in late 1986. In turn the naval clashes between the two countries in 1987 and 1988, and the shooting down of the Iranian airliner with the loss of 290 passengers in July 1988 have maintained the depth of anti-American feeling in Iran.

Such an accumulation of negative experiences will prove slow to overturn. There was virtually no hope of any substantive warming in relations during the remaining few weeks of the Reagan presidency. His caricature had become such a common sight that the regime could not afford to be seen to be doing business with him. However, George Bush, despite his membership of the Reagan administration, is not similarly tainted. He appears to have escaped most of the anti-Reagan criticism aired in Tehran over the preceding two years. On the US side, the continued incarceration in Lebanon of its nationals by Shi'a terrorist groups with links to Iran will restrict the pace at which the new US President can proceed. The release of the hostages is probably the *sine qua non* of a normalisation in relations, as far as Washington is concerned.

The obstacles involved in the improvement of Iranian-US relations and the slow speed at which they are developing provide short- and medium-term opportunities for the rest of the Western world. In its reconstruction efforts Iran will be looking for technology transfer, the like of which it can only obtain from Europe and Japan. In particular, states like West Germany and Japan, which continued to cultivate Iran even during its times of acute isolation, are well placed to profit. In relation to arms sales, Iran is likely to put the emphasis on diversified supply, in view of the difficulties experienced in the servicing of US equipment following the post-revolutionary break with Washington and the problems encountered in purchasing sophisticated weapons during the later stages of the

Gulf conflict. The diversification of arms procurement also has the added bonus of making it easier to play one supplier off against another, thus negotiating down the price. With an integral part of arms purchases being the presence of foreign personnel for training and maintenance purposes, the regime will probably be chary of making purchases from the US and the Soviet Union. Again Europe seems to be well placed.

Though the relationship between Iran and the US may only improve slowly, its potential, ironically enough, is greater than for Soviet-Iranian relations. This has much to do with geography. The logistics of external intervention by the US remain extremely difficult. Past US influence in Iran has been possible because of the economic, political or military presence already established. Starting now from a position of no contact at all, it would be easier for Iran to control the numbers of US citizens in the country and to monitor their activities. Furthermore, it is almost inconceivable that US troops would be landed in Iran other than in response to a Soviet invasion of the north of the country. As far as the Iranians are concerned the threat from the US can therefore be controlled, certainly more easily than that from the Soviet Union.

For the USSR current relations with Iran are better than those with the US. Unlike the US, the Soviet Union has an embassy in Tehran and consular representation in Isfahan. Both political and economic delegations make regular visits to Tehran from Moscow. Considerable potential exists for an expansion of oil and gas exports to the USSR. Natural gas exports were broken off at the time of the revolution, just as the second Iranian Gas Trunkline (IGAT 2) was under construction. During 1988 there were discussions about resuming deliveries via IGAT 1 (which operated successfully for nearly ten years) or converting the line to oil transportation. Both these developments seem unlikely. The Soviets could take Iranian gas, but they were forced to make arrangements for domestic supply to the southern republics and hence much of the rationale behind that trade has disappeared. As far as oil is concerned, the idea of allowing the USSR control over even a small portion of Iranian oil exports would be difficult for any government in Tehran to accept. In energy matters and in general, Moscow has skilfully used the economic domain as a way of maintaining regular contact without Tehran becoming nervous about the political content of such contacts.

The Soviet Union showed itself ready to court considerable unpopularity both in the Arab world and at the United Nations in the second half of 1987 by refusing to endorse an arms embargo resolution against Iran. Nevertheless, despite the not unfriendly level of existing relations, the potential for further improvement is not as great as with US-Iranian ties. The long common border alone makes Tehran extremely wary of Moscow. The proximity of the Soviet Union is arguably the reason for the measured rhetoric which emanates from Tehran directed at its superpower neighbour. In contrast, the fact that political abuse is heaped so readily on the Americans indicates the extent to which the regime does not feel threatened by the US.

Conclusion

The size and population of Iran ensures that, objectively speaking, Iran has the potential to emerge as the leading power in the Gulf. However, this potential cannot be realised until at least the second half of the next decade and possibly much later. The need to rebuild the country's oil industry, economy and armed forces all ensure that this state of affairs will continue. With Iran entering the 1990s at a military and economic disadvantage compared to Iraq, there are two reasons for anxiety. Firstly, there is the possibility that Baghdad might resort to the use of its military superiority in a limited engagement designed to extort political concessions from Iran, or to disrupt the reconstruction process. Secondly, there is the prospect that the seeds of a future conflict early in the next century may be sown as a result of Iraq using its superior power.

In the meantime, Iran is likely to pursue an essentially defensive diplomacy aimed at providing the breathing space with which to consolidate the domestic military and economic foundation. This is likely to be the case, almost regardless of the nature of the regime in power. While the future complexion of the government of Iran is difficult to predict, the probability of chronic instability and discontinuity after the death of Ayatollah Khomeini may have been exaggerated. Should national interest rather than pan-Islamism continue to provide the rationale for foreign policy-making, relations with the Gulf states and eventually the US will improve. Iranian support for opposition Shi'a groups will decrease correspondingly. The potential for an improvement in Soviet-Iranian relations, however, remains limited.

Notes

[1] The populations of the Arab Gulf states are as follows:

Iraq	16.3m
Bahrain	0.4m
Saudi Arabia	9.75m
Kuwait	1.7m
Oman	1.2m
Qatar	0.3m
UAE	1.6m
Total	31.25m

Source: Economist Intelligence Unit Country Report estimates.

[2] For example, Edward Mortimer gives the average age in Iran as 7.1 years, *Financial Times*, 16 August 1988.

[3] BBC Summary of World Broadcasts (SWB), Middle East, 14 September, 1988.

[4] Dilip Hiro, *Iran Under the Ayatollahs*, London, Routledge & Kegan Paul, 1985, p.308.

[5] James A. Bill, *The Eagle and the Lion*, New Haven, Conn., Yale U.P., 1988, p.408.

[6] Sir Anthony Parsons, *The Pride and the Fall*, London, Jonathan Cape, 1984.

[7] William A. Dorman and Mansour Farhang, *The US Press and Iran*, Berkeley, Univ. of California Press, 1987, p.13.

[8] BBC/SWB/ME, 20 July 1988.

[9] BBC/SWB/ME, 13 July 1988 quotes a letter from Iraqi foreign minister, Tariq Aziz, to the UN secretary-general in which he refers to such a resolution.

[10] BBC/SWB/ME, 22 July 1988.

[11] Ibid.

[12] BBC/SWB/ME, 20 July 1988.

[13] *The Independent*, 28 September 1988, said 30,000; the Associated Press, quoting US defence analysts, put the figure at 15,000 (*Jordan Times*, 30 August 1988).

[14] BBC/SWB/ME, 23 November 1987.

[15] *The International Herald Tribune,* 27 January 1988.

[16] *The Economist,* 12 February 1988.

[17] For example, see BBC/SWB/ME, 30 November 1987 in which it was claimed that 3 million volunteers for the front had already come forward, and 6 million were expected.

[18] That is, no large human wave attacks were to be launched. Instead, according to President Khamene'i in SWB 30 November, 1988 there would be continuous operations.

[19] *The International Herald Tribune* of 12 May 1988 gave the figure as 10. The London-based *Daily Telegraph* of 25 May 1988 said 200.

[20] *The Daily Telegraph*, 25 May 1988.

[21] *International Herald Tribune*, 12 May 1988.

[22] London-based *Guardian*, 19 March 1988.

[23] *The Guardian,* 25 July 1988. Kurdish expatriate sources claim that 6,000 civilians were killed in the attack.

[24] Literally, *velayat e-faqih* means the vice-regency of the jurist. Briefly, it is the 'concept of there being one man who is of such outstanding juridical and leadership quality that he may in effect act as vice-regent for the Prophet. This person will govern Islamic society and have the right to expect total obedience from all members in it'. Quoted from *Financial Times* survey of Iran on 1 April 1985.

[25] Hiro, *op. cit*. p.64.

[26] *The Guardian*, 5 September 1988.

[27] Spokesmen for the NLA quote the figure of 1,000 dead. Even if the figure is as low as this, it still represents a crippling blow to the force.

[28] Vahé Petrossian, writing in *The Guardian* of 5 September 1988 gives the figure as probably about 4,000.

[29] *The Guardian,* 5 September 1988.

[30] To quote Ayatollah Khomeini himself. See Haleh Afshar (ed.), *Iran: A Revolution in Turmoil*, London, Macmillan, 1985, p.22.

[31] Shahram Chubin and Charles Tripp, *Iran and Iraq at War*, London, I.B. Tauris, 1988, p.73.

[32] Shaul Bakhash, *The Reign of the Ayatollahs*, London, I.B. Tauris, 1985, p.17.

[33] Ibid. p.166.

[34] Ibid. p.194. The bazaar supported demonstrations in November, 1980 in favour of the foreign minister, Sadeq Qotbazadeh.

[35] EIU estimate in *Iran Country Report*, No. 2, 1987, p.2.

[36] Ibid. p.15.

[37] BCC/SWB/ME, 18 February 1988.

[38] See Baqer Moin, 'Questions of Guardianship in Iran', *Third World Quarterly*, Vol. 10, No. 1, January 1988, p.194.

[39] Afshar, *op. cit.*, p.221.

[40] BBC/SWB/ME, 22 August 1988.

[41] For example the State Prosecutor-General, Mohsen Khoiniha, the Minister of Security and Defence, Muhammad Reyshahri, and the deputy foreign minister for the Middle East, Husain Shaikholeslam.

[42] *Wall Street Journal*, 16 February 1988.

[43] Chubin and Tripp, *op. cit.*, p.38.

[44] With a handful of notable exceptions, such as the attempt on the life of the Amir of Kuwait in a suicide attack on 26 May 1985.

[45] For example those attending a liberation movement conference in Tehran in December 1987 included: Islamic Amal Organisation of Iraq; Hizbollah of Kuwait; Supreme Assembly of the Islamic Revolution of Iraq; Iraqi Islamic Dawa Party; Islamic Revolutionary Organisation of the Arabian Peninsula; Front for the Liberation of Bahrain; Hizbollah of Iraqi Kurdistan, BBC/SWB/ME, 4 December 1987.

[46] BBC/SWB/ME, 22 August 1988.

[47] For example, the US fleet retaliated against a missile attack on the *Sea Isle City* in Kuwaiti waters by destroying oil platforms in the Rostam field on 19 October 1987, and against the mining of the *USS Samuel B. Roberts* by striking the Nasr and Salman oil platforms on 18 April 1988.

[48] Chubin and Tripp, *op.cit.*, pp.294-5 and pp.303-4.

[49] BBC/SWB/ME, 13 September 1988

[50] *Dubai External Trade Statistics*, January 1988.

[51] BBC/SWB/ME, 26 July 1988.

[52] IRNA report in BBC/SWB/ME, 25 April and BBC/SWB/ME, 20 June 1988.

2. IRAQ

Introduction

Throughout the 1960s and 1970s the Western view of Iraq was almost uniformly and consistently negative - particularly in Britain and the United States. In the eyes of London and Washington Iraq had entered a dark age. The start of what Axelgard has called 'the persistent cynicism with which Western opinion and decision makers have viewed Iraqi politics'[1] undoubtedly began with the overthrow of the Hashimite regime in 1958. In the military coup d'etat of that year Nuri Sa'id, who had long dominated Iraqi politics and had largely done Britain's bidding, was removed. The end of the monarchy ushered in what has now been just over 30 years of Arab nationalist rule.

The various manifestations of republican government have all in different ways been extremely unpalatable to the West. First, there was General Abdul Karim Qasim, who, in his efforts to rival Nasser as the doyen of pan-Arabism, banged the drum of anti-imperialism ever more loudly. His reputed covetousness of Kuwait resulted in the amir of the recently independent country calling in British and then Arab League troops in 1961 to secure the integrity of the new state. Relations did not greatly improve with the arrival in power, following the reversal of the Ba'thist coup in 1963, of the Arif brothers. It was Abdul Rahman al-Arif who presided over the dispatch of what turned out to be a symbolic military presence[2] to the front in the 1967 war against Israel. Subsequently, diplomatic relations with the US were severed because of Washington's backing for the Jewish state. Then finally came the installation of the Ba'th Party in 1968, which began what has so far been 21 years of unbroken rule. Yet the experience of military takeovers in the previous decade has established in the West the view that Iraq is a 'coup-prone country'[3].

If anything, the Ba'th Party was even more anathema to the West than its predecessors. The zealous and uncompromising nature of the regime's ideology appeared directly to threaten Western interests, whether via its attempts to dominate the Gulf or through its anti-Western rhetoric. The close relationship which developed between Baghdad and Moscow, and the incorporation of the Communist Party into the Progressive National Front in 1973, appeared directly to threaten the free world.

Of special concern to the United States was the unyielding stand adopted by the Ba'th Party towards the Arab-Israeli question. If the party was hardline before it came to power[4] this continued at least rhetorically undiminished after the 1968 coup[5]. The style as well as the substance of Ba'thist policies were also deeply alien to the West. The secretiveness and conspiratorial nature of revolutionary politics in the republic, born of the party's experience of opposition and persecution, have always been profoundly unpalatable to the open societies of the West. Moreover, the repeated use by the Ba'th in both Iraq and Syria of insurrection as a means of seizing power was perceived as being deeply menacing. It suggested that even the most outwardly stable state in the region might be shaken from within by a putsch perpetrated by shadowy party cells.

The attitude towards Iraq in the Gulf states during this long period was similar to that of the West, only sharpened by the real security threat which Iraq posed. Its overbearing nature was manifested in many forms. To Kuwait, it appeared to be a threat to sovereignty and independence. Though there was no repetition of the 1961 crisis, the amirate was threatened in 1973 by the attack on the Kuwaiti border garrison at al-Samita. To the middle and lower Gulf, the threat was one of internal subversion. Ba'th Party cells, encouraged by Baghdad, were active during the 1970s in Bahrain. More widely, the smaller Gulf states had to contend with an Iraq that was desperate to assume a leadership role in the area, and was then frustrated and irascible when its ambitions were thwarted in the early 1970s. While the US was happy to entrust its regional interests to Iran and to a lesser degree Saudi Arabia through its 'twin pillars' strategy, the smaller Gulf states took some comfort from the fact that Iran helped to counterbalance the unwelcome interference of their domineering sister state.

This was the picture up until 1979. If this broad impression of Iraq endures in the West today, it is because of the intensity of this experience spread over 21 years, compounded by the style of Ba'th Party politics, rather than because of the continuity of policies over the last decade. Circumstances in the region changed irrevocably with the crumbling from within of the US' principal pillar, Iran. The Ba'th regime had already moved against the Iraqi Communist Party, executing 21 of its members and forcing some 3,000 others into exile when the Iranian revolution took place. Reliable, pro-Western, monarchical Iran was replaced by volatile, vindictive, revolutionary Iran. Rather than staunching the spread of the Ba'thist message, the strategic imperative, especially for the US, was suddenly to limit the boundaries of the Iranian brand of revolutionary Islam. Iraq became important in this context with the outbreak of the war in 1980. Once Iran appeared to have gained the upper hand in 1982, Iraq became increasingly crucial as the main bulwark against the forcible export of the new Iranian politico-religious creed. Recognition of the new positive role played by Iraq came with the resumption of diplomatic relations between Baghdad and Washington in 1984. By 1987 there had developed a *de facto* alliance between the two countries, mainly drawing on US naval superiority in the Gulf and Iraq's domination of the skies. This two-pronged pressure, together with the use of chemical weapons as a deterrent to war, ultimately forced the Iranians to accept Security Council Resolution 598.

If the war saw a confluence of interests between the US and Iraq *vis-à-vis* Iran, active co-operation was rendered easier by the sobering effect of the conflict on other areas of Iraqi Ba'thist policies. There has been a general winding down of the ideological zealotry which typified Ba'thism during the heyday of radicalism in the Arab world. In Iraq this has led to a greatly moderated position on the Arab-Israeli problem. Iraq is now content to support the mainstream leadership of the PLO and to back policies adopted by the organisation. In inter-Arab relations, Iraq has long made its peace with Jordan, and the two states now enjoy cordial ties. While Ba'thist Syria has emerged over the last ten years as the most bitter enemy, Iraq has established increasingly warm relations with the moderate states of the area, even with a post-Camp David Egypt.

Domestically, since February 1987 there has been a steady move away from the rigid socialism of state economic planning. The private sector is increasingly being given greater responsibility for economic activity, while the public sector is expected to be more efficient and commercial in outlook. Inevitably the regime remains an authoritarian one, capable of brutal and repressive deeds in order to retain the reins of power. Its use of chemical weapons against its Kurdish population in the north in 1988 has shocked public opinion in the West. Perhaps the extent of its internal relaxation will hold the key to the improvement of its image in the eyes of international public opinion in the post-war context. Whatever the developments on the domestic political front, it will be argued that in regional and international affairs Iraq is a different animal from that which prevailed 10, 20 or 30 years.

Strategic Interests

In the West there tends to be a fondness for discussing what are the strategic interests of the US, France, Britain and indeed the West as a whole in various parts of the world. Recently there has been a growing but cautious tendency to concede that the Soviet Union and the Eastern Bloc have legitimate national interests. However, there remains a reluctance to admit that Third World states have interests which are both of equal importance to the individual state and as legitimate as those espoused in the West. Rarely then are the national interests of individual regional states appraised, thereby leading to uncertainty over the objectives of states and incomprehension over the policies which individual states adopt.

In listing the national interests of the Iraqi state, it must be emphasised that these are not simply the quirky objectives of a self-aggrandising clique, but aims which are likely to endure regardless of the complexion of individual regimes. They are:

Ensuring the external security of the state. There can be no doubt that, after eight years of war, securing the state from external threats remains the chief objective of not only the Iraqi regime, but also the vast majority of its citizenry. In whatever guise the Iran-Iraq war might have begun, it soon grew into a struggle for the existence of the Iraqi state. Iran was committed to the liberation of the Shi'a areas in the south of the country. Its repeated attempts to capture Basra and its support for Kurdish groups in the north suggest that it was intent on nothing less than the dismantling of the Iraqi state. The establishment of the Supreme Assembly of the Islamic Revolution in Iraq (SAIRI) based in Tehran, and the agreement between a number of Kurdish factions which Iran forged in Tehran[6], also strongly imply that the Iranian vision may have extended to the dismemberment of the Iraqi state on at least sectarian and even perhaps ethnic lines.

Given the fact that Iran has consistently presented the greatest external threat to the contemporary Iraqi state, Iraq cannot afford to be indifferent to the potential re-emergence of a challenge from the east. Though the fighting may have subsided, the depth of the antipathy existing between the two regimes means that Iraq will not be able simply to take for granted the continuance of peace. This would certainly be the case in the event of a limited cessation of hostilities, whether formally or informally secured. However, even within the context of a formal negotiated settlement, should one occur, it will take some time for Baghdad to become convinced that Iran has abandoned its destructive intentions.

The establishment of a stable and lasting peace. The sort of peace which Iraq would like to see is a comprehensive, watertight and lasting one. Iran could participate with relatively less disruption in terms of economic reconstruction in an uneasy stand-off, or a formal ceasefire which falls short of a peace agreement, along the lines of the Syria-Israel model. The same is not true for Iraq, which has neither the manpower resources nor the strategic geographical depth of Iran. Anything short of a full and comprehensive peace would not allow Iraqi society to relax. The country would probably remain on a war footing in terms of the mobilisation and deployment of forces for fear of a sudden Iranian attack. The nature and quality of the peace is thus of much greater importance to Iraq than Iran. It is therefore ironic that Iraq has been the major cause of the slow progress in the peace talks which have taken place in Geneva and New York.

A secure peace would unlock a number of other key policy goals for Iraq. The wholesale return to civilian life of those conscripted into the army, some of whom have been serving for eight years, would have an immediate psychological effect on Iraqi society. It would help enable the population to put the trauma of the war behind it. Moreover, it would free further manpower resources to be used in the expansion of the economy likely to take place during a period of reconstruction. The switching of resources from the military to the civilian sector also applies to financial resources. At the beginning of 1988, even in the absence of a new Iranian offensive, the war was estimated to be costing the Iraqis some $500 m a month[7]. Although the immediate aftermath of the war will almost certainly not see the straight transfer of all military spending to the civilian sector, this figure gives some idea of the magnitude of the drain on the Iraqi economy represented by the hostilities.

Crucially then, much of the reconstruction process must be preceded by the achievement of a firm peace. Though the war has seen the relocation of much Iraqi industry away from the eastern border and deeper in the heartland of the country, there is little choice over the location of certain projects. The rebuilding of the port of Basra is essential because of its symbolic value in the war and its status as Iraq's second city. Iraq also has few options on the handling of its sea-borne cargoes because of its short coastline. In addition to rebuilding the port of Basra, Iraq will also want the unhindered use of the Shatt al-Arab and the Gulf for its shipping. The unimpeded use of the northern Gulf waters is also essential if it is to resume exporting its oil by sea. The relocation outside the mouth of the Shatt al-Arab of oil loading jetties and terminals which were destroyed early on in the conflict, is also an important consideration. None of these vital developments can be achieved without the establishment of security on the land front as well as at sea.

<u>Ensuring the security of its supply routes.</u> A lasting achievement of the war has been to diversify the supply routes to and from Iraq, thus making the country more strategically secure. Prior to the conflict, it was dependent on the long narrow waters of the Gulf for the passage of most of its imports and its exports of oil. Its only port facilities were Umm Qasr on the small stretch of coastline which borders the Gulf and, more importantly, Basra, at the top of the slender Shatt al-Arab. The route was a precarious one, and easy to close, as the war demonstrated. Iraq's

major oil pipeline was through Syria, and this was closed in April 1982. In the course of the war military supplies came in via ports in Kuwait and on the eastern shore of Saudi Arabia. Other supplies, such as foodstuffs and consumer goods, arrived via Aqaba in Jordan and Alexandretta in Turkey. The facilities at the former were expanded extensively to provide for the transit trade with Iraq.

For all its diversification, none of these supply routes is totally secure. Land and sea communications will continue to depend on the goodwill of other major regional powers, notably Turkey, Israel, Iran and Saudi Arabia, and also on Iraq's ability to manage its relations with these states. All that the policies of the last nine years have achieved has been to create alternative high-risk routes to the smaller number of high-risk routes which existed before the war. Consequently, Iraq will remain vulnerable to the closing of supply routes in the future, though it will be less dependent on individual routes, and thus less susceptible to pressure from individual states, than it has been in the past.

This strategic dependence means that Iraq, more than most states in the region, has an intrinsic interest in the policies both foreign and domestic adopted by other states. This is a function of the political geography of Iraq and of the region, rather than of the ideology of a particular regime. A more ready appreciation by other states of this fundamental interest on the part of Iraq would help to ensure that Iraqi motives are not misunderstood. It would also put outside powers in a better position to differentiate between Iraqi national interest and policies which might be construed as motivated purely by the desire to establish political hegemony over the region. The diversification of supply routes should at least enable Iraq to be less interventionist in its regional policy.

Maintaining the production and exportation of its oil. Iraq's prosperity and strength will continue to depend upon its ability to produce and export large quantities of crude oil. Prior to the war, the value of crude oil exports had risen rapidly to over $26 bn by 1980. The effect of the loss of its ability to export its oil production was shown in the early years of the conflict. Deprived of its sea and later its pipeline outlets, Iraq was reduced to exporting some 725,000 barrels a day in 1983, compared to 2.5 million barrels a day in 1980. That represented a fall in earnings of some $16 bn[8]. As a consequence of this, and allied to the need for increased military procurement, Iraq first drew down its considerable reserves and then ran up a national debt which, to the West alone, was estimated to be between $25 bn and $30 bn in 1988[9]. This was in spite of the fact that Saudi Arabia and Kuwait, which between them are believed to have extended loans worth some $25 bn to Iraq, were also helping to fund the war effort by selling crude oil from the neutral zone on Iraq's behalf.

In the early years of the war Iraq's oil exports were easily curtailed through the destruction of its facilities at the head of the Gulf and Syria's closure of its only pipeline in 1982. Some crude and refined products were transported by tanker truck through Jordan and Turkey, but this proved expensive and could not replace the volume lost. Iraq's position was strategically enhanced by the upgrading of one and the construction of a second oil pipeline through Turkey, with a combined capacity of 1.5 mn b/d, and the pumping of 0.5 mn b/d through the Petroline across Saudi Arabia to Yanbu. These new pipelines have not freed it from risk, however. As with its supply routes, diversification has simply transferred the risk. Instead of dependence on Syria and Turkey as alternatives to the sea route, Iraq is now dependent on Saudi Arabia and Turkey. Indeed the analogy with road transport supply routes is deceptive. There is a long time lag and great capital cost involved in the construction of pipelines compared with the utilisation of general port and road facilities. Notionally at least, then, Turkey and especially Saudi Arabia now have considerably more leverage over Iraq because of the pipeline factor than they had in the past. This also makes it more vulnerable to other regional actors. The Kurdish people, straddling the Iraqi-Turkish border, have at least the potential to disrupt exports. The location of the Saudi port of Yanbu, relatively close to the Israeli border, suddenly increases the risks of a direct Israeli attack aimed at Iraqi interests or the vulnerability of its outlet in the event of Israeli strikes against the Saudis.

Playing a leadership role in the Arab world. Since the foundation of the modern state, the Iraqi political leadership has tried to play a wider role in the Arab world. In particular this was true of the first King Faisal, Nuri Sa'id and Abdul Karim Qasim. Indeed, there is an intrinsic quality about Iraq's claim to leadership which transcends individual ruling regimes. Its demographic size makes it, along with Syria, the largest state in the Arab east. Its oil wealth, its much vaunted agricultural potential and (again along with Syria) the relative sophistication of its population reinforces this profile. The ideology of the Ba'th Party, which prescribes Arab unity as one of its trinity of tenets, underscores this inherent aspiration. It has been argued that regional ambition, dressed up in the desire to reincorporate the Arab province of Khuzistan into the wider Arab political community, was the driving motive behind its invasion of Iranian territory in September 1980. But Iraqi ambitions to play a leading role in the Arab east transcend individual regimes in Baghdad.

While post-war Iraq may be expected to continue to subscribe to these aspirations, it would be misleading to think that this situation will necessarily be destructive. The regime's overriding priority in the aftermath of the war will be with domestic reconstruction. In terms of foreign affairs, Baghdad can be expected to continue to look with suspicion to the east from whence the biggest threat will still be perceived as originating. Indeed, if the Iraqis continue to fear Iranian resurgence then they have a stake in not alienating the Gulf states. Moreover, they have in some respects learnt the technique of a more subtle leadership through cultural means, for instance. Thus, even during the war, they were able to maintain their leadership of the Arab Gulf in the province of education, by means of the establishment of a number of academic associations.

Domestic Politics

The war has dominated the internal political scene over the last nine years as well as external relations. Indeed, with the conflict having continued for so long, domestic politics operating within the framework of outside threats and hostilities has become the norm. Yet one must guard against the most common mistake of forecasters, to see the future as a projection of the present. With one chapter having closed on the Iran-Iraq war, the context in which domestic politics will take place in the future has changed. Similarly, one cannot simply look back to the pre-war period for a paradigm of Iraqi politics in peacetime. The experience of the war has been a catalyst with some profound effects.

The war and internal threats

Although one should be cautious in using the politics of an Iraq at war to conjecture about the politics of an Iraq at peace, there are some important conclusions to be drawn from the experiences of the war years. These relate to the domestic political challenges posed to the regime and the state by those two apparently most alienated of constituencies: the Kurds and the Shi'a. When pundits look at the likely threats to the leadership of Saddam Husain and the dominance of the Ba'th Party, the Kurds and the Shi'a routinely come to the fore. Yet, despite the sense of expectation that these communities might precipitate wholesale political change in Iraq, this has not been the case.

There can be no doubt that the war and the external strains imposed upon the Iraqi regime presented an extremely encouraging context in which opposition movements might operate. For the Kurds in the north, there was the very real prospect, spread over a number of years, that the Iraqi state would disintegrate. Indeed, during the later stages of the war the Iraqi army, concerned to conserve both men and material, effectively abandoned large areas of Kurdistan in the extreme north and north-east of the country where the remit of the state no longer applied. For the Shi'a in the south, there was the example of the self-conscious and proud Islamic revolution to the east. Also there was every likelihood that Iran, through success on the battlefield, would 'liberate' the areas of Iraq in which they lived and establish an independent state on their behalf.

Those who waited for the national liberation of the Kurdish people or the uprising of the Iraqi Shi'a were disappointed. To explain the non-occurrence of their predictions, many analysts sought simply to depict the situation in terms of the severe repression perpetrated by the Iraqi regime. It is certainly true that the Iraqi state resorted to brutal methods against the Kurds and the Shi'a opposition. Some of the Kurdish areas have suffered horribly. One has only to think of the resettlement policies in the north, involving the demolition[10] of many villages east of the Kirkuk-Mosul highway, and the use of chemical weapons against the inaccessible strongholds of the Kurdish resistance groups in the mountains. With respect to the Shi'a, activists have been imprisoned and killed, the execution of Ayatollah Muhammad Baqer al-Sadr in April 1980 being the best known case of the removal of an implacable opponent. However, brutality and repression alone certainly do not explain the widespread quietism prevalent amongst the Shi'a community throughout the war. Other factors must have been important.

First, there was often an error of analysis on the part of those who expected more of the Kurds and Shi'a in opposing the Iraqi regime. In the case of the Kurds, there is a temptation to romanticise. The Kurds are regarded as a self-conscious people frustrated by existing states in the desire and drive for national recognition. They are often viewed as the Palestinians of the east. However, this is erroneous. There are Kurdish intellectuals who couch their political analysis in the mainstream European nationalist mode, but they are very much in the minority, though arguably one which is growing. Rather it is kinship affiliation which continues to command far greater loyalty within the Kurdish areas. In other words, the Kurds of northern Iraq are as riven by internal kinship divisions as were the clans of the Scottish highlands in the eighteenth century. In political terms, these divisions are perpetuated at least in part by the leaders themselves, who retain command of their clients through traditional politics. It is no coincidence that when Mustafa Barzani, the grand old man of the Kurdish resistance in Iraq, died, he was succeeded by his son, Mas'ud.

Thus the Kurds remain riven and fragmented by clan and tribal politics. This both retards their ability to galvanise themselves into a definable, coherent ethno-linguistic group, and leaves them prey to divide-and-rule policies on the part of the state. The efficient way in which the Iraqis have set about exploiting these divisions is a second reason why the Kurds have been unable to take advantage of the opportunity presented by the war. The Iraqis managed to prevent a common front of Kurdish opposition factions from emerging until December 1986, not least by engaging the Patriotic Union (PUK) led by Jalal Talebani in discussions about increased autonomy in the northern areas. Even after Tehran was successful in forging this alliance of groupings, it certainly did not appeal to the whole Kurdish community. The 'tame Kurds' of the lowland areas had been wooed in the 1970s by the Iraqi Government through the provision of what were, for a highly centralised state, reasonably generous autonomy measures. Although security was massively increased in the lowland areas in the aftermath of the Tehran accord, much of the responsibility for security in the north remained with the *fursan*, or Kurdish militia force, which has stayed loyal to the Iraqi state.

There has also tended to be some faulty judgement in outside analysis of the Shi'a of Baghdad and southern Iraq. The fact that they subscribe to the same brand of Islam has led to the assumption in many quarters that the community is monolithic. Of course, the fact that they share the same confessional beliefs is a factor which tends to reinforce mutual identification. In the context of the strong magnet of Iran it is proper to examine to what extent they have felt the pull of the

ideological pole to the east. However, Shi'ism is only one of the elements whic
help mould the identity of the people in southern Iraq. A second common facto
is Arabism. More importantly, there is a tendency to exaggerate the broa
characteristics which people share, like Shi'ism and Arabism, and to undervalu
the importance of more potent claims on individuals, especially those of kinshi
loyalty. Even within Shi'ism there is a tendency to ignore the cleavages which exist
especially over the allegiance which practising Shi'a owe to the individua
mujtahid[11]. By not delving more deeply into the organisation of the Shi'a sect
the ideological and personal rivalries which exist within it have tended to b
overlooked.

The external perception of the Iraqi regime as purely repressive has als
blinkered observers to the subtle way in which it has sought to manage differen
political constituencies. The Iraqi state has been shrewd and efficient in backin
those factions among the Shi'a whose approach functions most in its own interest
The most obvious ideological cleavage which existed among the *ulama* wa
between the conservatives and the reformists over the extent to which ther
should be anti-government activity. As the latter gained ground in the late 1970
the Ba'th regime adopted a pragmatic reply. The *ulama* willing to support th
government were the beneficiaries of greater central resources, and more mone
was allocated for the adornment of Shi'a shrines. Important days in the Shi'
religious calender began to be openly acknowledged and celebrated, and Shi'ism
and Islam in general were recognised as important elements in Iraqi history an
culture[12]. There were also attempts to get Shi'a visibly represented at th
highest level, and some 40 per cent of the National Assembly, elected in 1980
belonged to the Shi'a variant of the Islamic faith. The practice of increasing th
number of Shi'a occupying visible senior positions continued in the 1980s. A
random sample of senior bureaucrats at or above the position of director-genera
or its equivalent showed 50 per cent of the posts to be occupied by Shi'a. A sampl
of 60 army officers at or above the rank of brigadier-general showed some 25 pe
cent to be Shi'a.

In addition to supporting the collaborators among the Shi'a religious leaders,
and recognising the importance of Shi'ism and Islam as belief systems for the Iraqi
people, the state moved ruthlessly to crush the organisation and leadership of the
Shi'a opposition. The execution of Sadr was probably the single most effective act
in this direction. As Mallat comments: 'When as-Sadr was killed, the momentum
that his charisma had generated was rapidly lost'[13]. Later the wholesale
imprisonment and execution of members of the Hakim family occurred as they
emerged as the leadership base for Shi'a opposition.

The period of the war shows that the coherence and power of the Kurdish and
Shi'a constituencies have been greatly exaggerated. Even in the context of the war,
when the state was at its most vulnerable, their ability to mount an effective
challenge to the regime was unexpectedly modest. Some concessions were won,

notably by the Shi'a community in terms of both patronage and resource allocation, but these appear to have been sufficient to neutralise the opposition rather than to encourage it.

In the context of peace the ability of either community to mount an effective challenge appears to have been further reduced. The dissident Kurds have already been the target of a concerted attempt by the state once again to extend its authority throughout the whole of the north. Chemical weapons have been used to rapid effect in cowing the areas of greatest opposition. The PUK have lost the best opportunity they had to extract further concessions on autonomy from the regime. They now appear to be divided over tactics, with their secretary-general, Jalal Talebani, favouring negotiation and the majority group in the union continuing to oppose talks. In the future, the Iraqis will remain vigilant to make sure that the Iranians do not continue to aid the Kurds in order to keep up the pressure on Baghdad, as they did in the early 1970s. Therefore, if there is to be a full and lasting peace, a clause about the non-interference in each other's internal affairs, such as was included in the Algiers Agreement of 1975, would seem certain to be a non-negotiable item on the Iraqi agenda.

The southern areas, which have been the scene of so much fighting, are likely to see much of the reconstruction effort devoted to their part of the country. Though there will be a tendency to site new industries away from what has been the war front for fear of any future outbreak of hostilities, there appears little doubt that the port city of Basra will be rebuilt. Its importance as a trade conduit is crucial, given the small coastline which Iraq possesses. More important still is its symbolic value. Basra is Iraq's second city and the jewel for which the Iranians vainly sacrificed so much. Its regeneration as the epitome of Iraqi resolve to keep the invader at bay will be paramount, and Basra could well go down in Iraqi school history textbooks as the contemporary Qadisiya[14]. The reconstruction of a city which was so badly pounded by artillery is also vital if its inhabitants are to be attracted back and rewarded for the discomfort which they stoically endured.

Perhaps of greater importance will be the demoralising effect which the end of the war will have on those who harboured fantasies of the establishment of an Islamic republic in the south of the country. Khomeini and his revolution have been humbled by the way in which the war has ended. The myth of the inevitability of the spread of the revolution has been exposed, as has the divinity of the crusade to export its political creed to adjacent areas. It is extremely unlikely that Iran would contemplate the extradition of the SAIRI leaders to Iraq but, in the context of an overall peace, the activities of the organisation will probably be drastically curtailed. However, it must also be likely that Iran, keen to end its international isolation and the long and debilitating war, will be prepared to scale down its assistance for Islamic liberation groups, not least in Iraq.

Post-war political challenges

The disarray of the extra-systemic opposition inside Iraq largely precludes the possibility of an uprising or a takeover of power by a cadre grouping. Yet the regime is still vulnerable. With the creation of the Saddam phenomenon, a personality cult of extravagant proportions even by the standards of the developing world, the system is vulnerable to the removal of the 'struggler-leader' at its core. To a small, disorganised, yet committed opposition the path of assassination holds the most promising, indeed perhaps the only, potential. Moreover, the Ba'th Party, with its pure but narrow membership base of some 30,000, is also vulnerable to a concerted campaign of assassination. The 14 months prior to the invasion of Iran saw a militant Shi'a group, al-Dawa, mount just such a campaign. The most recent large-scale attacks credited to the organisation were at Dujail in July 1982 and Baquba in September 1987.

By its very nature, assassination as a mechanism of political change does not require widespread organisation or legitimacy, and therefore cannot be ruled out. In a country until recently at war the materials of the assassin are relatively freely available. Nor does the assassin have to be ideologically orientated. In a society where a refined sense of honour and loyalty to kinship groups run concurrently, personal motives, such as the death of a relative in 'Saddam's war' or at the hands of the security forces, could provide the necessary motive. Neither does a potential assassin have to be drawn from one particular ethnic or confessional group. Alternatively, assassinations can be planned and executed by outsiders or those acting on behalf of other regimes to which Saddam Husain is anathema, like Iran or Syria. It is the vital ingredient of opportunity which has done most to thwart the potential assassin.

Though Saddam Husain has been projected as the personification of the Ba'th Party, it would be misleading to dismiss the party as a body of political significance in the possible event of his death. It is true that the senior ranks of the party have been purged of personalities capable of rivalling the President. Indeed, it is difficult to envisage any of the current membership of either the Revolutionary Command Council or the party Regional Command filling his shoes. Speculation has occasionally lighted on Taha Yassin Ramadan, the first Deputy Prime Minister, as a rival to the President. While Ramadan is the only conceivable though still unlikely single alternative leader in either body, it is erroneous to present him as a competitor to Saddam Husain.

In spite of the absence of rivals at the pinnacle of the Ba'th, one should not underestimate the extent to which the party, which has been in power for over 20 years, has permeated Sunni Arab society in particular in Iraq. Though it retains a tight cadre organisation, it has between 500,000 and one million 'camp followers'. Furthermore, many more Iraqis, especially in the rural towns, have benefitted from the fruits of the oil wealth distributed as development spending by the Ba'thist regime, especially in the 1970s. Certainly in the event of the death of the President it would be essential for the senior leadership of the party to close ranks. The close association of the other members of the RCC and Regional Command and indeed the whole party with Saddam means that in the event of the regime collapsing they would be unable to distance themselves from the experience of two decades of Ba'th rule. It would therefore be in the interests of the rest of the senior Ba'thists, as well as those in and associated with the party, to hang on to power, most probably in some form of collective leadership. Indeed, it is noteworthy that when rumours abounded in 1982 about the President being replaced, it was a triumvirate of senior party figures who were suggested as the future leadership[15]. Were such a scenario to unfold, outside observers might be surprised to discover that the Ba'th Party is better consolidated and enjoys greater legitimacy than many had tended to assume.

A key constituency in the transition of power would be the senior officers of the armed forces. Although the President has made strenuous efforts to ensure that his generals do not build up independent support among the troops they command, with a vacuum existing at its centre, the military would be important in the aftermath of the President's demise, not least because of the instruments of force which it holds. In order to give greater backbone to a collective leadership, prominent members of the military might well be co-opted into the RCC. This would represent a reassertion of the importance of Ba'thists in the military, partly at least at the expense of the civilian party. In other words, there could be a return to the 1968 period when, in the wake of the coup d'etat by Ba'thist officers, the military wing of the party was far more powerful than it is today.

Even without the demise of the President, in the post-war context the armed forces could play a more important role in the politics of the state. This is in part a function of the military's enhanced position in the aftermath of hostilities. The fact that the Iraqi army successfully kept the massed Iranian land forces at bay for six years will have boosted its reputation and raised professional pride. The army has also played an interesting role in terms of nation-building. In spite of some desertions, it emerged as a unifying institution capable of galvanising the Shi'a for instance, whose loyalty was considered to be dubious, in the defence of the state. In fact the Shi'a provided well over half the non-officer army corps in the struggle against the Iranians.

However, the army's contribution was not simply a defensive one. The storming victories achieved at the very end of the conflict in recapturing the Fao Peninsula and Shalamcheh meant that it could claim to have been instrumental in breaking the Iranian military machine. Moreover, the army can claim these achievements more or less for itself. Following the disastrous defeat at Mehran in 1986, responsibility for which was attributed to the President, Saddam Husain relinquished a measure of his control over military decisions in favour of his generals[16]. The battlefield victories at the beginning of the summer of 1988 can therefore be claimed with conviction by the generals and by the army as a whole.

The fillip to the armed forces provided by the outcome of the war is not limited to the army, but extends also to the air force. The war witnessed an expansion of the resources of the Iraqi air force. For instance, its combat aircraft increased from 339 in 1979[17] to some 500 in 1988[18]. The air force also played an important part in bringing the war to an end. Its role in disrupting the Iranian economy did much to undermine Tehran's war effort. The attacks against economic targets, such as power stations, which resulted in extensive electricity cuts, also helped to grind down Iranian morale. The sorties against Kharg Island increased the cost to Iran of exporting its oil, while the long-distance raids on the Larak and Sirri oil loading platforms in the lower Gulf showed that Iran's extensive geography was not necessarily a guarantee of security.

The success of the armed forces in the war does not alone make them a threat to Saddam Husain. The continuing suspicion of Iran is likely to mean that senior military personnel will remain preoccupied by possible external military challenges. There are also the northern Kurdish areas which have to be maintained under the authority of the state. There may also be an increase in tension with Syria. In addition, the senior officer corps tends to have a Sunni majority, and the President has cultivated close links with them. Saddam Husain also regularly switches his generals round, to ensure that a personal bond of loyalty does not develop between individual troops and their senior commmanding officers. The regular changes in the position of chief of staff, regardless of ability or success, illustrate this point. Moreover, it is questionable to what extent one can refer to the armed forces as a coherent, unitary body. The expansion of the elite Republican Guard could be important in this respect. Three years ago it consisted of eight brigades. By 1988 there were 26.

Despite these caveats, there is a state of unease between Saddam Husain and the army which dates back to the 1970s. This is partly born out of the President's sense of insecurity in never having served in the military. This tension continued with the war as the President faced the quandary of needing the army to defend the country, but not wanting it to become too powerful. Now he faces an army which is self-confident and may feel it has earned a greater say in the running of the country. While a direct challenge to the regime remains highly unlikely, the President may have to co-opt military leaders more directly into the decision-making structures, especially if the economic reconstruction process falls short of expectations.

A loosening of the grip?

During the 1980s Iraqi society has been controlled with a relentless grip. The authoritarian characteristics of the regime predate 1980, although the outbreak of the war gave the authorities the opportunity to justify and extend their control over society. The ending of the war will in turn remove a central rationale for this rigid control, especially in the Sunni Arab heartland of the republic. This situation will be greatly enhanced if a formal peace is concluded between Iran and Iraq, especially if it contains clauses forbidding the interference of the former adversaries in each other's internal affairs. Consequently, it can be argued that in a climate of peace the maintenance of such a rigid grip on Iraqi society will be harder to justify. Of course the ending of the war is not in itself a panacea. The extent to which armed resistance to the state persists among certain Kurdish factions and among underground Shi'a groups will have some bearing on the general political climate. Nevertheless, the transition from war to peace is still a major one. Furthermore, it is likely that there will be a general yearning for a perceptible increase in fundamental personal freedom. This demand for a relaxation of society will come principally from the intelligentsia.

There has been a large expansion in the ranks of the educated middle classes over the last generation. For instance, the number of public sector college students rose from 8,568 in 1958-9 to 75,270 in 1975-6, while those attending state secondary schools jumped from 73,911 to 499,113 over the same period[19]. Before the war the Ba'thist leadership was at least aware of the rising demands for greater political participation in the country. The creation of the National Assembly fulfilled a decade-old promise made by the party to establish a broader base of legitimacy for the regime. The elections held in June 1980 - the first national poll since 1958 - were a small, tentative step in the direction of involving the people in the political process.

Now the dissatisfaction with the present state of affairs seems likely to grow. Unlike Iran, where most of the fighting was done by the poorer sections of society and the middle classes were able to buy exemptions for their sons, the middle classes in Iraq played their part in the war effort. Doctors and engineers were obliged to serve continuously in the army, because of the need for their skills. The bureaucracy was no safe haven from the war, with the President requiring that even senior officials should spend time at the front. Educated women filled the gaps left in the public sector, and were able to rise to hold senior positions. With the middle classes having proved their commitment to the state, it will be politic for the regime to respond to their calls for some increase in both political participation and personal freedoms.

Signs that this will be the direction in which things will go were suggested even in the context of the war. Well before the ceasefire the second National Assembly had played a greater role, especially with reference to the political accountability of the government, than its predecessor. The most visible change in its activities was its authorisation to criticise the performance of government officials including, at least in theory, members of the RCC and the President himself. To date the assembly has contented itself with the relatively easy target of the

Ministry of Health. As a result of its inquiries, the Minister, Sadiq Hamid Allous
and a significant number of lower ranking officials were sacked in May 1988[20
An article by a leading expert analysing possible changes quoted a source i
Baghdad as speculating that as a next step the assembly might take responsibilit
for approving the appointment of cabinet ministers[21].

A further small but significant sign of the regime's desire to adapt to the wishe
of the intelligentsia has been the development of an 'open wall' system. At a
educational institutions in the country, from primary schools to universities,
place has been designated where students can post notices expressing grievance
To date the subjects of such posters have been socially rather than politicall
orientated; for example the excessive costs of getting married. However, ther
are indications that the authorities are taking this type of expression seriousl
with the National Assembly reportedly having taken up the particular issue c
marriage costs[22].

While these incremental steps in the direction of relaxing controls ar
noteworthy, they only hint at the direction which events may take after the wa
They are certainly not irrevocable. Indeed, if, despite the ceasefire, a high degre
of internal subversion were to break out, then this trend might be arreste
However, reservations about the likely extent of the relaxation in the political an
intellectual spheres are more than compensated by profound and fast movin
economic reforms.

The initial details of economic reform were announced in a speech by Sadda
Husain in February 1987. Since then, further changes in the regulation of th
economy and in the organisation of the public sector have regularly been made
Essentially, what is in effect a 'commercialisation policy'[23] falls into two parts
First, the improvement of the performance of the state-owned economy, with
whole stratum of administration, the state enterprises, being removed. Mor
important perhaps has been the move to give greater responsibility for th
economic success of individual factories to their managers. In return for a 'fre
hand', the managers must turn their projects into commercially viabl
ventures[24].

To facilitate this, the dead hand of the Ba'th Party has largely been remove
from the factory floor. In the past the party was heavily involved in the work place
and would routinely take the side of the employee against the management. No
the interference of party functionaries in industrial relations has been largel
curtailed. This thrust has been backed up by other fundamental changes. In th
state sector trade unions were abolished overnight by the reclassification of a
blue-collar workers as white-collar workers. Furthermore, the liberal labour la
was abolished in March 1987.

The new regulations have done much to undermine the security which state employees have enjoyed in the past. Moreover, in the drive to reduce costs, workers have lost a whole range of fringe benefits, such as free transport to work. More stringent working practices have also been introduced, such as longer working hours. The incentive to the workforce has been a compulsory 25 per cent share in the profits, or losses, made by the state enterprises relative to work targets.

Since February 1987, when this new trend was established, there is no sign of any reversal or let-up in the implementation of the policy. There have been anxieties about the policy on the part of the workers. Senior bureaucrats, who have lost some of their administrative fiefdoms, are understood to be unhappy. But the policy is unlikely to be reversed. First, because it has gone too far. Secondly, because it appears to have been successful particularly with regard to boosting productivity. Indeed, it was the increase in output of the state armaments factory in an experiment run by its head and Saddam Husain's son-in-law, Husain Kamil al-Majid, that originally convinced the President of the wisdom of the changes.

The second part of the reform is to encourage the private sector to play a more creative role and adopt a higher investment profile. In implementing this strategy the state has been selling off some existing public sector concerns, notably in the agricultural sector. It has also sought to encourage co-operatives and the establishment of mixed enterprises, with private investors forming a majority on the governing boards regardless of the size of the stake. There are even plans to mobilise the savings of individuals into productive sectors with the incentive of guaranteed returns[25].

The government has also sought to encourage domestic industries to develop export markets. The principal incentive in this respect has been to permit businessmen to keep 50 per cent of all the foreign currency which they earn on such exports. This foreign currency can in turn be used to import any commodities, regardless of whether they are linked to the exporting enterprise. For a country where foreign currency is extremely hard to secure and where imported luxury goods are sold at high mark-ups, the new incentive promises bumper profits for the exporter-cum-importer.

In addition to making the Iraqi economy more efficient and more responsive to the country's needs, the new policies appear strongly to favour those with considerable accumulations of private capital. More particularly, they seem to present opportunities for the old established industrial families to entrench their positions of wealth. Not only does this represent a considerable dilution of Ba'thist theory and practice in relation to the implementation of socialism, but it also appears to be aimed at broadening the base of the regime to include some of the older and wealthier elites.

Foreign Relations

Even though it has emerged from the Gulf conflict in a stronger than expected position, there can be no doubt that Iraq has suffered thoroughly debilitating effects which have taken a great toll on the country's human, financial and material resources. A great yearning exists for the transfer of priorities from the military to the civilian sector in so far as this is possible without prejudicing the security requirements of the state. Consequently, in the immediate future there is an aggregate desire for resources to be given to the internal rather than the external sphere, and for peace rather than war.

The Gulf

The Gulf region will continue to be the primary focus of Iraqi foreign policy well into the future. The experience of the long and bitter war with Iran and the importance of the area for Iraqi national interests demand it. In particular the Gulf waterway will return to its position of pre-eminent concern as the bulk of Iraq's supplies and oil exports resume the use of this route. The importance of Saudi Arabia for the export of Iraqi oil by pipeline also ensures a continued preoccupation with the region. However, Iran will remain the principal focus of Iraqi concern.

Revolutionary Iran continues to be regarded as the chief threat to the security and integrity of the Iraqi state. Baghdad's philosophical perception of Iran is revealing with regard to the depth of this suspicion. Iran is derided as not possessing a modern ideology based on rationality. Instead, the regime in Tehran is scorned as being grounded on the idea of an Imam who disappeared centuries ago. A regime so squarely based on the traditions of the Imams for its legitimacy, it is argued, cannot give up its aim of liberating a town like Samara, where three of the Imams are buried, or Najaf and Karbala, which house the tombs of the important religious figures of Ali and Husain. In other words the ideological base of the Iranian regime, as perceived from Baghdad, makes the occupation of Iraqi soil an enduring objective. The view of the Iraqi leadership is that this will continue as long as the *vilayat-e faqih* remains the fountain of authority in Iran. Consequently, the Iraqi leadership does not regard Iran as a rational nation-state with which it can deal on the basis of any sort of mutual trust. Even the Shah's regime is contrasted favourably. The point is made that it was possible to have a 'civilised' discussion with a representative of the Shah. If an understanding was reached its implementation could at least be relied upon.

This view of Iran is compounded by the fact that Iran was felt to be at its lowest ebb at the point when it had been 'squeezed' so tightly that it was forced to accept SCR 598. Tehran did not accept an end to the war out of conviction, but simply as a device to gain a respite from Iraqi attacks and to end its isolation from the international community. With the tourniquet loosened, from the time of the ceasefire onwards Iran could only get stronger, at least in financial, demographic and military terms, and Iraq's recurring nightmare is that it will persist in its greed for Iraqi soil, and that the time will come when it will resume hostilities.

With Iran beginning to enjoy an improving material position and possessing a regime which may continue to harbour long-term designs on Iraq, Baghdad must continue to be on its guard. It must be vigilant against attempts to destabilise it internally, and threaten it externally. The internal threat will necessitate the retention of an internal security apparatus operating in particular in those areas whose loyalty to the state is most suspect. It will also require the increased channelling of development expenditures in these directions as a top priority. Thus, it is the south which will be the focus of the major reconstruction programme for reasons other than purely the desire to repair war damage.

The external threat demands that Iraq should continue to maintain a large standing army on its eastern border. This means that demobilisation can only be partial, and male Iraqis will probably be required to perform periodic terms of military service. Suspicion of Iran will also require the allocation of significant proportions of public funds for the continued procurement of sophisticated defence and punitive strike equipment. With the rapid growth of the Iranian population and with fears that the ending of Iran's diplomatic isolation may enable it to buy advanced weaponry, there is an even greater imperative for Baghdad to retain its qualitative advantage over the Iranian military. In the light of Iran's experience in the war, it is vital for Iraq to possess the sort of deterrent that will discourage further Iranian military adventures in the medium and long term. The limp response of the international community to repeated revelations of the use of chemical weapons by Iraq is unlikely to discourage the mass stockpiling of such weapons as part of this strategic deterrent.

Iraq's basic and enduring fear of Iran makes it all the more important for Baghdad not to alienate the Gulf states. The pursuit of abstract ideals of Arab unity or the implementation of policies aimed at undermining moderate Arab governments are therefore likely to be eschewed. In the course of the hostilities, Iraq came to recognise the importance of Kuwait, Saudi Arabia and Bahrain in particular to its war effort. Kuwait and Saudi Arabia have more recently grown in importance as routes for the export of Iraqi oil. Moreover, the impression persists that Riyadh's oil policy, which helped to give Iraq a free hand within OPEC while forcing down Iran's oil revenues through a depressed spot-market price, was instrumental in applying the sort of pressure which forced Tehran to accept the ceasefire.

For its part, Bahrain emerged as a key land base for US naval facilities during the height of the tanker war in 1987 and 1988. The presence of the US fleet was vital in supporting the Gulf states, and to a lesser degree helping to increase the pressure on Iran. All in all, it would be extremely unwise for Baghdad to return to the radical pan-Arab rhetoric which made its neighbours to the south so uncomfortable in the past. It would be even less clever to go back to fostering the sort of internal subversion which was aimed in the 1970s at undermining the conservative regimes in these states. Similarly, it remains improbable that Iraq will seek to reawaken its territorial dispute with Kuwait. Such a move would instantly alienate the Gulf states as well as ending the regional and international support which Iraq has enjoyed in the course of the conflict. For all these reasons, Iraq may be viewed as a supporter of the existing power constellation in relation to the Arab Gulf.

Instead, its attempts to play a leadership role in the Gulf are more likely to be couched in positive language. Baghdad will no doubt look to consolidate its leadership of the area in cultural terms with the establishment of more pan-Gulf organisations based in Baghdad, and headed by Iraqis. There may also be further impetus for economic bodies and industrial and service companies to be established based on the six GCC states plus Iraq. Ultimately, there may be an attempt to join the GCC, especially if these other seven-state experiments in co-operation get off the ground. Such an approach would have to wait until a stable peace had been achieved in the area, for if tension persisted it would be easier for the GCC states to argue against Iraq's admission on the basis of security. If a stable peace is achieved, however, it would prove extremely difficult for the individual GCC members to resist Iraq's overtures, given its credentials as an Arab Gulf country. Any capitulation on its admission would come about in spite of the clear and united opposition which the GCC states privately share on the undesirability of Iraq's membership. Should the GCC be expanded to include Iraq there is a distinct possibility that Oman would withdraw from the Council.

While the substance of Iraq's policies may be heavily diluted compared with those of a decade ago, the style of foreign relations is less likely to be the subject of review. Saddam Husain has seen that his overbearing way of dealing with the leaders of the Arab Gulf states in private discussions and negotiations has worked very effectively. As a corollary, the implicit threat of some of Iraq's older and less palatable objectives, such as territorial readjustment or support for opposition forces, will remain in being in case the core support for Iraq should appear to be substantively weakening.

The Levant

Iraq's continued preoccupation with the threat from Iran and its considerable national interests in the Gulf make it unlikely that there will be any rapid or substantive change in its policies towards the states to its west. Anxieties have been expressed that now that the hostilities have ended Iraq can afford to return to its old radical policies and turn its back on those predominantly moderate states which gave it support during the trying times of the war. More particularly worries have been expressed in Jordan that Iraq will care less about the Kingdom's friendship now that the needs of the war no longer require cordial relations. Some have spotted the potential for renewed tensions with Egypt as both Baghdad and Cairo seek to compete for the leadership of the Arab world.

Such concerns are very much anchored in the regional perspective of the late 1970s. Circumstances ten years on have certainly changed. Iraq is now much more conscious of its own strategic vulnerability. With direct supply routes through the Gulf and Syria having been cut off, Baghdad has been keen to diversify into alternatives as much as possible. The Red Sea port of Aqaba in Jordan proved one such extra supply route. Indeed, the Iraqis actually contributed to the expansion of Aqaba port as well as making funds available for the repair of the Desert Highway from the Red Sea to Iraq. Having developed this supply line as an important strategic alternative, they would be foolish to lose it.

However, the close links with Jordan are likely to persist for other reasons. King Husain was perhaps the most consistent and visible supporter of Iraq throughout the war. For instance, even when he was trying to bring the Iraqi and Syrian leaders together at the Arab summit in Amman in November 1987, it was always on the basis of the need for concessions from the latter over their support for Iran. With the Jordanian economy stagnating and with chronic current account deficits, the prospects of a reconstruction boom in Iraq has been greeted as a possible salvation for the Kingdom. In addition to the prospects of exporting its skilled manpower, notably its unemployed engineering graduates, and the excess capacity in its ailing construction industry, Jordan also hopes that Iraq will repay the $600m credit line as well as resuming some of its aid payments, which were suspended in the early days of the war. Such a situation makes Jordan more dependent on Iraq and thus increases the opportunity for leverage over Amman. It would therefore be pointless for Iraq to turn its back on a state which is growing increasingly receptive to its influence. Baghdad's goodwill gift to the Kingdom of 204 tanks and personnel carriers in mid-August 1988[26] suggests that it will not do so. In fact, both states may even feel it is in their interests to join together in a confederal relationship in the future.

Although there is more potential for competition between Iraq and Egypt, there is no intrinsic reason why there should be a return to the situation of acute rivalry which prevailed between Qasim and Nasser. With Iraq continuing to be wary of Iran to the east and Syria to its rear, and with Israel's antipathy for it apparently unabated, it is not in its interests to alienate the other major actor in the region. Moreover, the fact that Egypt is still excluded from the Arab League reduces the degree to which Cairo could rival Iraq on the Arab stage. And even if it were to be formally readmitted to the Arab fold the fact that Egypt is the only Arab state to have signed a bilateral peace treaty with Israel still reduces its standing in intra-Arab politics. It also effectively removes it from whole areas of Arab concern, such as the threat from Israel and the best way to confront it. If the potential for inter-state rivalry is sometimes over-emphasised, there is even less cause for friction from inter-personal competition. While Saddam Husain may well entertain some of the ambitions to lead the region that characterised Abdul Karim Qasim, Husni Mubarak is no Gamal Abdul Nasser. The leadership style of Mubarak, with its low-key and bureaucratic approach to politics, presents no obvious clash with the strong personality and charisma of Saddam.

One of the recent quiet successes of Iraqi diplomacy has been the way in which it has extended relations with the Palestine Liberation Organisation. These have been improving ever since the estrangement of the PLO from Iran in the early 1980s. The PLO's breach with Jordan in February 1986 along with the insecurity of both the organisation and the Tunisian Government as a result of the Israeli raid on Tunis in 1985 prompted an organisational reorientation towards Baghdad. The Iraqi Government helped facilitate this by making the PLO's presence no longer conditional on the diplomatic policies it pursued, in contrast to the interference on the part of Jordan and especially Syria. In fact since 1981 the Iraqi position on the Palestinian problem has been not to oppose any solution which is acceptable to the PLO[27].

This ostentatiously hands-off policy towards the PLO was logical enough in time of war. Providing facilities for the organisation was an astute way of keeping up Baghdad's importance in the context of the Arab-Israeli dispute. It was also a clever way of linking Iraq with what continues to be the paramount issue in the Arab world, namely the plight of the Palestinians. In the aftermath of the ceasefire Iraq's position could alter, but is unlikely to precipitate any major shift. Now that the war has ended Baghdad can no longer play a passive role in relation to the Palestinian issue if it has serious aspirations to leadership in the Arab world. Therefore, one would expect to see Iraq adopting a more challenging posture. However, this is likely to be in support of the mainstream PLO rather than aimed at undermining it. The organisation's chairman, Yassir Arafat, remains almost as much an anathema to President Hafez al-Assad of Syria as Saddam Husain does. They are therefore natural allies against a Syria which has vigorously attempted to undermine both of them. A more strident advocacy of the Palestinian cause by Iraq at multilateral forums such as the Non-Aligned Movement would be one relatively uncontroversial way of Baghdad's raising its profile on the issue.

Indeed, it is Syria, certainly as long as it continues to be ruled by President Assad, which remains the enduring enemy of fellow Ba'thist Iraq in the Arab world. The barriers which have kept the two states apart are unchanged: two leaders with an implacable personal animosity; two branches of the same fervent ideology competing to be the one legitimate upholder of the faith; two adjacent states of similar size that are the natural rivals for pre-eminence in the eastern Arab world. Compounding this fundamental enmity is Iraq's vindictiveness over the support rendered by Syria to Iran during the war. While the state of Iraq will be happy to see the standing of Syria further eroded, Saddam Husain has personal reasons for seeing the pressure mount on his counterpart in Damascus. This increased friction is likely, however, to continue to be played out by proxy. Iraq will maintain its diplomatic support for Syria's Christian adversaries in Lebanon, and will continue to supply arms to them. Both the Syrian and the Iraqi regimes could also explore ways of trying to destabilise one another internally. Any military posturing, let alone armed clashes, are remote, however. The Iraqi people are almost as weary of war as their counterparts in Iran. Swapping one conflict for a second against a large well armed force would not be popular in Iraq.

The probable future foreign policy of Iraq is good news for Israel. Iraq will remain preoccupied with Iran, while its first area of regional interest will be the Gulf. To the west, it is likely to continue its relationship with the moderate states of Egypt and Jordan which either have made or seek peace with Israel. Moreover, Iraq will almost certainly remain as implacably opposed to Syria (the most threatening Arab state because of its location and its policy of strategic parity) as it has ever been over the last decade. The only sense in which a resurgent Iraq might be unpalatable to Israel is in relation to its encouragement of the PLO. Suffice it to say that it is the moderate wing of the organisation with which Iraq is most closely linked. And this is a PLO which has adopted an encouragingly moderate strategy in the context of the *intifada*, most importantly through the acceptance of UN Resolution 242 at the meeting of the Palestine National Council in Algiers in November 1988. While Israel may look with suspicion at Baghdad, Iraq is unlikely to be a major obstacle to the possible conclusion of a comprehensive Arab-Israeli peace.

The superpowers

In a world where there is still a tendency to slot states into either a US or a Soviet sphere of influence, Iraq is usually placed in the latter. However, this is vastly to exaggerate the relationship which exists with the Soviet Union. Although Iraq has in the past been dependent on Moscow for much of its military supplies, it has succeeded in partially diversifying its purchases of advanced military equipment. In particular, it has established a relationship with France which has supplied Baghdad with Mirage aircraft and helicopters. The main impetus to seek other sources of military procurement came from the unreliability of Soviet supplies and Moscow's linkage of arms supplies to its overall interests in the area. The Iraqis felt let down during the civil war with the Kurds in 1974-5 when the Soviet authorities increased Baghdad's problems by witholding arms supplies. The Soviet Union cut its supplies of arms to Iraq to a trickle in 1980 following the invasion of Iran. Moscow felt aggrieved that its complex manoeuvres aimed at wooing the new revolutionary regime in Tehran had been jeopardised by the Iraqi use of Soviet weaponry in such a way. It is believed that the disruption of supplies remained in force until 1982 when the Soviet Union suddenly became more concerned at the possibility of an Iraqi defeat.

These two episodes have made Iraq suspicious of the Soviet Union's reliabili as an arms supplier, and disappointed at the general quality of the relationshi In 1982, for example, Saddam Husain criticised the Iraqi-Soviet friendship trea for failing to measure up to Iraqi expectations. The frustrations have not only bee confined to the issue of arms supplies. Iraq has felt that the Soviet Union has als let it down on the diplomatic stage. In November 1987 the Iraqi Foreign Ministe Tariq Aziz, spoke openly of his 'dissatisfaction' at Moscow's refusal to back th UN Security Council in follow-up measures against Iran for not accepting SC 598. By the same token, however, the leverage of Moscow over Baghdad, whethe in the diplomatic or the military domain, has proved to be marginal, as the Sovi Union's ignorance of the decision to move across the border into Iran in 198 illustrates.

Iraq's need for reliable supplies of advanced military hardware and th vascillation of the Soviet Union in the area in the past suggests that Iraq woul like further to diversify its purchases. The possibility of the US making inroad into this market appears limited, owing to the probable blocking of any such sale by the pro-Israel lobby in Washington. Indeed, in 1987 the Reagan administratio even refused an Iraqi request for the relatively innocuous C-130 transpo aircraft[28]. However, the successes of France in arms sales to Iraq would sugge that Europe could play an expanding role in this respect.

In spite of the blind spot on weapons supply, US-Iraqi relations have improve considerably over the last few years. The re-establishment of diplomatic relation in 1984 was an important milestone, and signalled Washington's approval c Baghdad as a government with which it was happy to deal. This formal resumptio of ties grew into informal co-operation in the military and diplomatic spheres i confronting Iran. Iraq and the US found themselves with a congruence of purpos in trying to contain the Iranian military machine in 1987 and 1988. At the UN, th US was the most vocal advocate of back-up measures - most visibly an arm embargo - to force Iran to accept SCR 598. Although the Iraqis are inevitabl suspicious that an unseemly scramble between the two superpowers to court post-peace Iran may take place, the extent of the co-operation between the U and Iraq has warmed relations considerably.

The general atmosphere of goodwill between the two states, together with th changing nature of Iraqi economic policy, presents considerable opportunitie for the US. The encouragement of the private sector, which has already bee clearly signalled, is an area where US expertise could be harnessed. The attempt to re-invigorate the agricultural areas also hold promise for US exporters an consultants. One commentator has fixed on the agricultural sector as the are which is the most hopeful in terms of US assistance. He has advocated that Iraq access to agricultural credits should be expanded and repayment term relaxed[29].

Action on such a non-controversial front could hold many benefits for the US. There is the obvious one of expanding its trading and consulting profile in a large market, which is to be the focus of a considerable investment in reconstruction. This broader foundation of interaction would inevitably expand the basis of political understanding. This in turn would help the US to encourage Iraq to maintain its moderate, non-confrontational policies in regional affairs. Ultimately, this would favour all those states which found Iraq's foreign policy of the 1970s most distressing: the main allies of the US in the area, namely Israel and the smaller Gulf states.

Conclusion

For the duration of the war, the chief threats to state and regime respectively in Iraq were perceived as coming from opposition groups within the Kurdish and Shi'a communities. The failure of these to capitalise upon the war shows the relatively weak challenge which they pose. In the context of the aftermath of hostilities, it is other areas of the domestic constituency that the regime must seek to conciliate; amongst these are the military, the intelligentsia and the private sector. Increased economic and to some extent political liberalisation may be expected as a response to the aspirations of the intellectual and entrepreneurial middle classes. The Ba'thist government's relatively egalitarian policies towards resource allocation applied to the process of reconstruction should ensure that the benefits are spread throughout the society.

The geography of Iraq and its core national interests necessitate that Baghdad pursue an active regional foreign policy. This should not be confused with the ideological objective of spreading the Ba'thist revolution which was a high priority in the 1970s. In fact in the late 1980s Iraq can best be described as a *status quo* power, which seeks to work with rather than subvert the existing Arab Gulf governments. The continued perception of Iran as being the greatest external threat to Iraq will ensure that Iraq remains most preoccupied by the Gulf in general and its eastern flank in particular. In the absence of a comprehensive and durable peace, Baghdad will be obliged to maintain a high degree of preparedness in the event of the resurgence of any future threat from Iran. The preoccupation with Iran will ensure that Iraq does not pursue a policy of confrontation with either Israel or Syria. In the Arab world in general, Iraq will continue to co-operate closely with the moderate states, especially Jordan over which it will be able to exert increasing influence. Potential for enhanced US-Iraq economic and political relations is limited. The utility of the relationship with the Soviet Union will therefore continue to be perceived, ensuring that close ties are maintained.

Notes

[1] Frederick W. Axelgard, *A New Iraq?*, New York, Praeger, 1988, p.48.

[2] Hanna Batatu, *The Old Social Classes and the Revolutionary Movements of Iraq*, Princeton, New Jersey, Princeton U.P. 1978, pp.1,065-6.

[3] *The Economist*, 20 October 1984.

[4] For example, in its rejection of Security Council Resolution 242 in November 1967.

[5] For example, its rejection of the Rogers Plan. See Batatu, op. cit., p.1,096.

[6] An agreement was concluded between the two main Iraqi Kurdish groups, the KDP and the PUK, in Tehran on 8 November 1986. For the main themes of the accord see the *Kurdish Observer*, February, 1987.

[7] *Mideast Markets*, 7 March 1988.

[8] Figures taken from Economist Intelligence Unit Country Profile of Iraq.

[9] Ibid.

[10] According to Kurdish expatriate sources some 3,500 villages and hamlets in Iraq have been razed.

[11] Someone qualified to exercise *ijtihad*, that is 'independent reasoning from first principles'. See Patrick Bannerman, *Islam in Perspective*, London, Routledge, 1988, pp.246-7.

[12] Shibli Mallat, 'Iraq' in Shireen Hunter (ed.), *The Politics of Islamic Revivalism*, Bloomington, Indiana U.P., 1988, p.79.

[13] Ibid. p.82.

[14] The scene of a famous victory won by the Arabs over the Persians. It was repeatedly invoked by President Saddam Husain, both to portray the Iran-Iraq war as a wider conflict between Arabs and Persians, and to instil confidence in victory into his people.

[15] They were Taha Yassin Ramadan, Tariq Aziz, Foreign Minister and Sa'doun Hamadi, the current (1988) Minister of State for Foreign Affairs. See Dilip Hiro, *Iran Under the Ayatollahs*, London, Routledge & Kegan Paul, 1985, p.211.

[16] *The Sunday Times*, 17 August 1986.

[17] *The Military Balance 1979-1980*, London, IISS, 1979, p.40.

[18] *The Military Balance 1988-1989*, London, IISS, 1988, p.102.

[19] Batatu, op. cit., p.1,120.

[20] SWB/BBC/ME Summary of World Broadcasts, 24 May 1988.

[21] Fred Axelgard in *Middle East International,* No. 328, 24 June 1988, p.20.

[22] Ibid.

[23] *Mideast Markets*, 7 March 1988.

[24] Interview with under-secretary at Ministry of Finance, 11 February 1988.

[25] *Mideast Markets*, 7 March 1988.

[26] *Jordan Times,* 17 August 1988. Also *Council for the Advancement of Arab British Understanding Bulletin,* Vol. 14, No. 17, 15 September 1988.

[27] Axelgard, *A New Iraq?* op. cit., p.83.

[28] Ibid. p.101.

[29] Ibid.

3. THE GULF CO-OPERATION COUNCIL STATES

Introduction

The Gulf Co-operation Council (GCC), consisting of six Arab Gulf states: Saudi Arabia, Kuwait, Oman, the United Arab Emirates, Bahrain, and Qatar, came into being in May 1981. The impetus for the establishment of the organisation was concern about Gulf security. The GCC was formed against a backdrop of three events which had destabilised the area and which in different ways appeared to threaten the smaller states of the littoral Gulf. These were: the Islamic revolution in Iran; the Soviet invasion of Afghanistan; and the start of the Iran-Iraq war. Therefore, there existed a powerful rationale for these states to group together in order to try to achieve strength and resilience to political change.

The timing of the establishment of the organisation, in the wake of the outbreak of the war, was significant as it permitted the smaller Gulf states to exclude the two main regional powers - Iran and Iraq. This was no small achievement as both states have historically claimed a right to play a broader security role in the region. Indeed, during the meetings on Gulf security in the mid 1970s both Iran and Iraq had participated[1]. In order to justify the exclusion of these two big neighbours, the GCC charter referred to the 'special relations and similarities' existing among the Council's members. In so doing, the group identified itself as a discrete body consisting of Gulf Arab states with traditional forms of government.

There was also an attempt to make the new organisation less obviously concerned with security affairs. Consequently, the charter sought to give the GCC a broader brief mentioning plans to draw up joint regulations covering 'the economy, finance, education, culture, social affairs, health, communications, information, passports and nationality, travel, transport, trade, customs, haulage and legal and legislative affairs'[2]. However, the suspicion has always been that, as one commentator put it, these aspects have simply been 'window dressing'[3].

In its preoccupation with security matters, the GCC has addressed itself to both external and internal threats. Its main response to the former has been the establishment of a 'rapid deployment force' with the strength of a 'brigade plus'[4]. This force is located at the King Khalid base at Hafr al-Batin in the north-east of Saudi Arabia, and draws its forces from all six members of the GCC including Oman. However, because of demographic realities and Oman's reluctance to be identified too readily with the force, the personnel is provided mainly by Saudi Arabia. Kuwait provides two battalions, and Oman a company.

Writing about the security policies of the GCC, Laura Guazzone maintains that individually and collectively the GCC states cannot hope to match Iran and Iraq militarily. The philosophy underlying the security policy of the six states is 'to build their collective and national assets so as to provide a military deterrent sufficient to make any direct confrontation as costly as possible to their adversaries'[5]. However, this aim even from a collective perspective is some way from being realised. Certainly, the GCC has no navy to speak of and plainly did not have the capacity to protect its shipping from the attacks by Iran during the Gulf conflict. To date, one can only agree with a senior army officer from one of the GCC member states who regards the combined force as purely 'symbolic'.

With regard to internal security, the GCC has arguably been rather more successful. There has been extensive co-operation amongst the intelligence services of the various member states[6]. Kuwait has been reluctant to endorse some of the internal security measures, notably those pertaining to hot pursuit and extradition.

Successes

Evaluating the performance of the GCC is extremely difficult. First, there is the problem of measurement. How exactly can the influence of the GCC on states external to the organisation be measured? Similarly, how can the general impact of the organisation on the politics of the area be gauged? Secondly, there is a problem of comparison. It is certainly difficult to compare the performance of the GCC with that of the individual states prior to its creation. It is also impossible to say how the six states, operating separately, would have fared during the turbulent period of the later stages of the Gulf conflict had they not been part of the Council. Thirdly, there is the problem of delineating the actions and influence of the organisation compared with those of the individual states. What can be termed GCC actions and what are the actions and influence of individual states within it?

The difficulties involved in measurement and comparison inevitably mean that judgements on the effectiveness of the organisation are both subjective and rather crude. Generally speaking, the conclusion of many analysts and commentators has been that the GCC is an ineffective and weak body. There is also a tendency to dwell on its failures. It is accused of being unable to avert the outbreak of embarrassing disputes among its members, let alone protecting the constituents from external attack. This may be the fashionable view, but it certainly does not give the whole picture. As with the UN and other multilateral organisations there is a tendency to be impatient with the performance of the GCC, and to ascribe unrealistic expectations to it.

Yet one is left with the feeling that there is more to the GCC. After all, if it is supposed to be such a tepid success, why do the member states themselves persevere with it? Perhaps a clue to its continued importance lies in the fact that the Council looks set to continue functioning as an integral part of the politics of the region in the 1990s. Given that the GCC was created primarily as a response to the Iran-Iraq war, a diminution of its status in the aftermath of the ceasefire might be expected. The fact that the GCC is to continue indefinitely suggests that its members perceive it to have some efficacy.

Broadly speaking, five ways in which the GCC has proved itself to be a subtle success might be identified:

1. It provides a forum within which the six states can interact. Many valuable functions are provided by the chance for heads of state to meet annually and foreign ministers to get together more often. It enables the leading officials of different states to get to know each other. It also gives them the opportunity fully to discuss different issues and to understand the perspectives of the other members. It can also provide a good opportunity to identify potential friction between members and for the others to launch mediation efforts in good time.

2. It permits a channel to be opened with regional and international powers, using the state with the best relevant relations as the conduit. This area has proved a particular strength of the organisation since its establishment, and has been of notable use both on the regional and international levels. Regarding the former, perhaps its most important use has been in keeping a channel of communication open with Iran when the war was raging. This has been done both formally and informally. Following the GCC summit in Riyadh in December 1987, the UAE retained the chairmanship of the committee charged with maintaining a dialogue with Iran. The following month, a UAE delegation travelled to Tehran[7] to explore the possibility of a meeting between the GCC and Iran. On an informal level, during this time Oman also continued to keep contacts open with Iran.

Individual states also provide good relations with various actors in the broader international sphere. The most obvious example is Kuwait, which has long had diplomatic relations with both the USSR and China. Kuwait's relations with the Soviet Union are positively good, to the extent that the Emirate felt that it could approach Moscow in late 1986 with the idea of re-registering some of its merchant fleet under the hammer and sickle. Indeed, when the GCC was first formed, Kuwait was the only member to have formal relations with Moscow. Even now, when other members have established relations, they do not necessarily have resident representation. The Soviet ambassador accredited to Muscat for instance, resides in Aden. All the individual Gulf states have diplomatic representation with the major states of the West. However, that does not mean that the quality of the representation[8] and the flow of information is equally good. The relationship between Washington and Riyadh, for example, is likely to be more sophisticated than that with the smaller states. King Fahd's personal partisanship in favour of the US would tend to suggest that Saudi Arabia is best placed to gain the ear of the US.

3. It gives the member states greater clout within other multilateral organisations. Chief among these is the League of Arab States. The emergence of the primacy of the Gulf conflict over the Palestinian question, as the leading issue preoccupying the Arab world in 1987, was evidence of this. The convening of an Arab summit in November 1987 in Amman was the pinnacle of this policy. The meeting was the first summit when the Palestine question had not topped the agenda. The GCC also gained an extra bonus from the meeting. In spite of the attendance of Iran's only Arab ally Syria, the final communiqué was tough and direct in its condemnation of Tehran. It spoke unequivocally of 'Iranian aggression'. When the GCC summit took place a month later, its communique was noticeably milder. It did not mention Iran by name. The convening of an Arab summit favourable to its perspective on the Gulf conflict thus enabled the GCC both to have a strongly worded condemnation of Iran, and to project itself as a milder, more measured body. The whole manoeuvre was no doubt intended to convince the Iranians that the Gulf states were not antipathetic towards them.

4. Nevertheless, the GCC enables the smaller states to adopt more uncompromising policies. Were the smaller Gulf states not linked in such an organisation, they would have little choice but to bend before whichever breeze was blowing hardest in the area. However, the existence of the Council enables them both to endorse hard-hitting resolutions and to be 'invisible' in that they are issued in the name of the GCC, rather than of individual members. Qatar in particular has found this aspect of the GCC especially attractive.

5. It has helped perpetuate the political *status quo* in five member states and the seven constituent parts of the UAE. The existence of the GCC has been a force for continuity in relation to the regimes in its various parts. More specifically it has also been an important stabilising factor in the continuation of individual

rulers in power, in spite of the destabilising events happening in the Gulf. Of course it is impossible to know if any of the various regimes and rulers would have been toppled or come under more substantive pressure had the GCC not been created. However, in the context of the Gulf conflict the latter would seem to be a reasonable assumption. The only example of a serious attempt at a palace coup during the first six years of the GCC's existence was the attempt to unseat the ruler of Sharjah, Shaikh Sultan, by his brother, Shaikh Abdul Aziz. Indeed, Sultan was temporarily deposed. The fact that he was restored some days later following feverish behind-the-scenes efforts including other members of both the UAE and the GCC suggests that agreement on a broad principle may have emerged over the undesirability of changes in leadership by such means. Certainly, all the GCC heads of state have a vested interest in stamping out a practice which was rife in the smaller emirates in the past.

Intrinsic Weaknesses

The GCC continues to be riddled with a number of weaknesses, which are intrinsic to the organisation. These effectively prevent it from emerging as a strong, unified and coherent block capable of joining Iran and Iraq as a third major force in the area. These inherent weaknesses may be characterised as springing from the geopolitics, historical experience and domestic composition of the member states. As such, they are immutable.

Geopolitics of the area. The GCC occupies a large land area. Its members are spread out in a linear pattern rather than being clustered together. This means they do not all share common borders. Over 500 miles separates the northern-most member of the organisation, Kuwait, from the nearest point of the most southerly, Oman[9]. This spatial diffuseness gives the various member states differing perspectives on the politics of the region.

Certainly the proximity of the Iran-Iraq war to Kuwait made the conflict of more pressing importance to the Emirate in its early stages. It was only when the war expanded into a Gulf conflict that the states of the lower Gulf began to feel exposed to its spillover effects. The recognition by all the Gulf states of the immediacy of the threat from the conflict in turn prompted varying interpretations and prescriptions. These came to the fore more earnestly towards the end of the conflict. For instance, at a GCC meeting in the early summer of 1987 Kuwait requested that GCC troops be stationed in the Emirate as a symbol of the other five's commitment to the integrity of the Kuwaiti state. The proposal was turned down because the other members deemed it a greater priority not to antagonise Iran further.

These differences in perspective can of course be exaggerated. In spite of the distance between them, both Kuwait and Oman, as well as their fellow GCC members, have many important interests in common. One such is the reduction of friction in the Strait of Hormuz. For Kuwait, and also Qatar, the Strait is the main outlet for their oil exports. Though Oman's oil exports do not have to pass through this narrow waterway, the Strait is important as the one point where Muscat's interests potentially rub against those of Iran. The whole of the Strait falls within the territorial waters of Iran and Oman[10], and each has responsibility for guaranteeing safe passage for shipping. When tensions rose and Iran, as in the past, began attacking selected shipping, Oman was in danger of being pulled into the conflict. Nevertheless, the differences of interest between Oman and the northern Gulf states are also illustrated in such episodes. The Omani solution to Iranian incursions into its territorial waters after November 1987 was to do nothing. While this avoided a possible confrontation between Iran and Oman, it left the shipping of Oman's fellow GCC members prey to the navy of the Islamic Revolutionary Guard Corps.

One of the clearest ways in which the difference in perspective, born of political geography, manifests itself is in the perception of where external threats lie. Of course the differentiation between external and internal threats can be misleading. For Bahrain, for instance, the external threat from Iran has to be considered alongside the existence of a large and in part potentially-alienated internal Shi'a population. This linkage is also important where significant numbers of fifth columnists have been recruited by an outside power. Again Bahrain is a good example, especially in relation to Iraq. In the 1970s the pan-Arabist Ba'th regime in Iraq recruited extensively in the Arab world. In the Gulf, Bahrainis proved particularly responsive. Though the cells which exist on the island are largely dormant, their reactivation remains a possibility. Similarly, pan-Islamist Iran has tried to recruit among the Shi'a communities and the 'oppressed' in the Gulf since the advent of the revolution. Bahrain proved to some extent fertile ground, though the potency of the threat from this quarter appears to have been greatly exaggerated.

Turning to conventional external threats, there is a tendency for analysts to differentiate between the northern states of Saudi Arabia, Kuwait and Bahrain, and the southern states of Qatar, the UAE and Oman. As a rough and ready method of categorisation, this division has many uses. However, the drawing of arbitrary lines inevitably erases the subtleties of individual states' foreign policies. Moreover, it glosses over the extent to which the major states of the GCC, Saudi Arabia, Kuwait and Oman, have markedly different external security preoccupations.

Saudi Arabia must certainly be viewed separately. It is the largest member state, uniquely possessing a Red Sea border and with credible aspirations to leadership within the Arab world. As a major actor on a stage greater than the Gulf, the potential threats to Saudi Arabia are not confined to the Gulf. The acquisition by Riyadh of long-range Chinese surface-to-surface missiles has most importantly brought the state of Israel within range. Saudi Arabia now joins the list of states that have to brace themselves for a potential pre-emptive strike from Israel.

Oman, positioned as it is mainly outside the Gulf, has to take a broader view of regional and particularly peninsula politics. In particular, historical and ideological tensions make the People's Democratic Republic of Yemen a major threat to Muscat. Oman's recent tradition of enmity with South Yemen also makes it conscious of Marxist Ethiopia in the horn of Africa, and attempts to extend the influence of radical politics in the area. For the other four GCC states, the external threats are concentrated in the Gulf. The location of the UAE, jutting out towards Iran, makes it, along with Oman, especially conscious of the proximity of Iran.

Kuwait is located next to Iraq, which has historically harboured expansionist visions at the Emirate's expense. Because of its concern about Iraq's ambitions, Kuwait has developed a foreign policy which seeks to maximise independence of action, while appeasing Baghdad over its core interests. One writer characterised this relationship by using the analogy of Kuwait as a 'pilot fish'[11]. Its relationship with Baghdad is, therefore, substantively different from that of its five fellow GCC members. Saudi Arabia tends to perceive Iraq as one of two powerful, potentially threatening regional forces. The other smaller Gulf states tend to identify three overbearing powers in the region, with Saudi Arabia being added to Iran and Iraq.

The importance of the differing perceptions of external threats is not exclusively applicable to the regional actors. A linkage is sometimes made between the actions of regional states and those of the superpowers. This is most vividly seen in the case of Oman. Its location on the eastern border of South Yemen colours Muscat's perceptions of Moscow, even at a time of moderation and restraint in Soviet foreign policy. Oman has long had a strained relationship with the Soviet-allied South Yemen. During the 1970s, Aden gave material and moral encouragement to the rebellion in Dhofar, in western Oman. Since the suppression of the revolt, relations have improved between the two neighbours. The civil war in South Yemen in January 1986 reawakened Omani concern. While inter-governmental relations are reasonable, the worry in Muscat is the extent to which the Aden Government wields real power in the country. With political stability in South Yemen thought to be fragile, the military in particular appears to operate largely autonomously, especially in the border areas. The location of an unstable and latently hostile state on its flank helps shape the Omani view of the area. It seems to be more than usually concerned about Soviet intentions in the region, especially holding to somewhat outdated views that Moscow seeks a warm water port on the Gulf.

Fear of domination. When they agreed to the formation of the GCC, there was some anxiety on the part of the five smaller states that they might become dominated by Saudi Arabia, which was the natural leader of the Council by almost any criterion. It was the largest in terms of size, and the only state to have common border with each of the other GCC member states. Its locatior bordering on the Red Sea as well as the Gulf, also gave it a strategic depth wit regard to supply routes which the other members, with the exception of Omar did not share. Saudi Arabia also had the largest population, being nearly six time as large as the next most populous state, Kuwait. Its oil wealth made it the mos prosperous state on aggregate in the GCC, while the leadership role which it ha played within OPEC meant that Riyadh already had an acknowledged diplomati role. If any state had the potential to transform the GCC into a formidabl regional actor, both diplomatically and militarily, it was Saudi Arabia.

Though they sought the security of GCC membership, the other five were no happy that the organisation should grow into a vehicle for Saudi aggrandisment The initial impetus to join had been to reduce the vulnerability of the individua states to the spillover effects of the Iran-Iraq war. The smaller states did not wan to escape domination by Iran only to be dominated by Saudi Arabia, a state which owing to its proximity, had the ability to exert greater influence at will over them Consequently, there has always existed within the GCC an ambivalence on th part of the smaller states. On the one hand, the external threat was a powerfu rationale for closer co-operation in all fields within the organisation. On the other the worry was that this would result in a loss of individual sovereignty to Saud Arabia and the GCC secretariat, which was somewhat unfortunately located i Riyadh. A tension, therefore, has existed between the two preoccupations, wit the degree of the external threat determining the extent to which one or the othe has been in the ascendant.

On the whole, the smaller states have proved clever in eluding the paternalis grasp of the Saudis. It also appears that Saudi Arabia has not been overl heavy-handed in trying to make the others submit to its leadership. The presenc of Kuwait, with its liberal institutions and tradition of individual freedoms, acted as an opposite pole of attraction in the early years of the GCC. However, even when the amir dissolved the Kuwaiti National Assembly and direct press censorship was introduced, Saudi Arabia's lead was only fitfully followed. The smaller states were willing to make certain concessions to Riyadh by way of appeasement. For instance, Saudi Arabia, in the name of the GCC, was allowed to protest to the European Community over its protectionist policy with regard to petrochemicals, which was depriving Saudi Arabia of potential export markets. Oman, however, was careful to reassure its European friends that bilateral interests would not suffer over what was essentially a Saudi complaint.

During this period, Saudi Arabia has come to exert more influence over one of the states within the GCC, Bahrain, with which it is now linked by a causeway. This causeway is strategically important, being wide enough for tanks to use, should there be a repetition of the aborted coup attempt in 1981. However, the chief motive for the construction of the causeway was not to bring the island state more fully under the tutelage of the Saudis. In fact the link has proved to be a useful outlet for domestic pressures within Saudi Arabia. Members of the Shi'a community in the eastern province are permitted to travel to Bahrain to celebrate the religious event of *ashura*. Saudis also use Bahrain as a recreational outlet, and there has apparently been no pressure on Manama to tighten its alcohol laws, as was feared prior to the opening of the causeway.

During the whole of its existence, the GCC has managed to avoid being dominated by any one party owing to the studious efforts to keep the structure of the organisation a loose one. Rather, it has remained an organisation within which all its members were able to 'continue to breathe'. Even with the strong impetus to closer co-operation provided by the war, the GCC has continued as a loose 'co-ordination of convenience', as one observer put it, rather than an emerging alliance. The advent of peace will certainly not result in greater centralisation within the GCC.

Historical disputes. Forging co-ordination within the GCC will continue to prove difficult because of the historical tensions which exist among the member states. These frictions vary in intensity. In general they help to maintain the inter-state suspicions which pervade and thereby handicap the work of the organisation. Three examples will illustrate the underlying mistrust which characterises the politics of the GCC. The examples chosen have been included because they occurred or re-surfaced recently.

The first and highly visible example was the clash between Bahrain and Qatar over the Fasht al-Dibal, little more than a sandbank near the disputed Hawar Islands. This occurred in April 1986, while the Iran-Iraq war continued to rage and the credibility of the GCC was most needed. The affair itself was one of the most recent episodes in a long chapter of friction and political competition between the ruling families of the two states, which had manifested itself very clearly during, for example, the negotiations for the establishment of the United Arab Emirates in the late 1960s and early 1970s. The Bahrain-Qatar division was superimposed on to and helped accentuate the Dubai-Abu Dhabi cleavage. Consequently, while Bahrain was allied to Abu Dhabi in the negotiations, Qatar and Dubai supported each other.

The 1986 incident was sparked off by Bahrain's attempted construction of a military harbour at Fasht al-Dibal. The island Emirate claimed to have GCC approval for the project, which was designed to monitor Iranian attacks on shipping down the Gulf's west coast[12]. Doha has long been incensed by the fact that Bahrain claims the Hawar Islands, which are located just off the western coast of Qatar. The Qataris were quick to resort to force. The Qatari military was dispatched to re-take the Fasht al-Dibal and to arrest the contractors working on the project. By the time the GCC formally intervened to ensure that the dispute did not deteriorate further[13], the affair had already become a public embarrassment, and stark evidence of the latent antipathy existing between the two states.

The second example relates to the enduring uneasiness between Oman and Saudi Arabia. The rapid expansion of the Saudi state in the first half of the century was of general concern on the peninsula. Many states lost territory as a result of this dynamic, notably Kuwait. Oman and Abu Dhabi nearly suffered as a result of Saudi efforts to take over the Buraimi Oasis to the north-east of Oman in 1952. The event caused great outrage in Oman and Sultan Sa'id bin Taimur assembled the tribes with a view to marching on the Oasis. Saudi forces occupied the Oasis for three years until 1955, but were then obliged largely by British pressure to withdraw. The Saudis as a result took to aiding the Imamate rebellion based on the Jabal Akhdar in northern Oman in the mid-1950s, and by such means kept up the pressure on Oman. The Buraimi Oasis incident is still remembered vividly, especially among the older members of Omani society, and is presented as a symbol of Saudi expansionism.

The issue of boundaries has re-surfaced periodically ever since. The last example of a border dispute between the two was in 1985. The respective heads of state, Sultan Qaboos and King Fahd, appeared to have cleared the air during a *téte-à-téte* at the GCC summit in November 1986 in Abu Dhabi. However, the Saudis have not been willing to reduce the pressure on the Omanis. Perhaps one of the most irksome instances of this has been the decision of the Saudi Interior Minister, Prince Nayif bin Abdul Aziz, to retain as an adviser the man who led the original Saudi incursion into the Oasis, Turki bin Ataisham. Saudi Arabia has thus chosen to keep up the psychological pressure on Oman, even though it has inevitably been at the expense of promoting partnership within the GCC.

The third example relates to the situation within the United Arab Emirates. Although the UAE is a sovereign state, its constituents have a marked (and recently growing) degree of autonomy. In many ways the UAE could be portrayed as a microcosm of the GCC. Two large Emirates in the shape of Abu Dhabi and Dubai dominate the federation, though Sharjah is large enough to have and pursue its own interests. The rest of the UAE's members are small and relatively weak components, with limited room for political manouevre.

The attempted coup in Sharjah in June 1987 was in part a product of the competition between Abu Dhabi and Dubai within the federation. Abu Dhabi supported the claims of Shaikh Abdul Aziz, who ousted his brother, Sultan, ostensibly because of his spendthrift domestic policies. Dubai was not prepared to see Sultan, with whom it enjoyed warm ties, removed. An eventual compromise resulted in the reinstatement of Sultan as the amir. The ruler of Abu Dhabi and the president of the UAE, Shaikh Zayid, was unhappy with the arrangement, especially as Sultan proved to be tardy in implementing the other parts of the compromise. Almost a year later Shaikh Zayid returned from a trip to Egypt with a former ruler of Sharjah, Shaikh Saqr bin Sultan, who had ruled the Emirate between 1951 and 1965 when he had been deposed by the present ruler's elder brother, Khalid. However, in 1972 Shaikh Saqr, who is a cousin of the present incumbent, returned and assassinated his successor. Following a period in prison, Shaikh Saqr has lived in exile in Egypt ever since[14].

Shaikh Zayid's decision to bring back Shaikh Saqr was not of great relevance in itself. Now an old man, Shaikh Saqr is unlikely to present a challenge to the current ruler. What the action did do was to deliver a gratuitous insult to Shaikh Sultan, and to confirm in his own mind that Shaikh Zayid was unhappy with his return to power. It thus may be expected to increase his suspicions of Abu Dhabi and its ruler, and of the apparatus of the federation which is based in Abu Dhabi. In other words, while contributing nothing in terms of the power balance in the Emirates, the return of Shaikh Saqr was likely to exacerbate disunity and the fractious nature of politics in the area.

Internal Stability in the Gulf States

The 1980s have seen the strategists once again worrying about the internal stability of the Gulf states. Indeed, the formation of the GCC was partly a response to the perceived threat from within. Real concern about domestic stability grew in the wake of the Iranian revolution. It was feared that those Gulf states with significant local Shi'a populations would suffer from growing instability as these communities became more self-confident and even began to emulate their co-religionists in Iran. The majority of the GCC states have sizeable Shi'a populations, namely Kuwait, Saudi Arabia, Bahrain and the UAE. There can be little doubt that chronic unrest in these states would also profoundly have affected a state like Qatar, even though it does not have a significant Shi'a population.

The fear of militant Shi'a-inspired unrest was great. The reality of the situation was much more subdued. There were some spontaneous sympathy demonstrations in the immediate aftermath of the Iranian revolution, especially in Saudi Arabia. There was a more calculated though far from successful attempt to overthrow the existing regime by Shi'a conspirators in Bahrain in 1981. There were periodic examples of some Shi'a groups resorting to the use of political violence, with varying degrees of success. There were regular incidents of minor bombings in Kuwait in the mid 1980s, for example, and a handful of more notorious attacks, one of which nearly claimed the life of the amir. Despite the existence of these occurrences, the political challenge posed by radical Shi'ism has been much more muted and far less well orchestrated than many would have imagined at the beginning of the decade.

There were two main explanations for the continuing quietude of the Gulf Arab Shi'a. Firstly, as in Iraq, other patterns of identity and affiliation proved to be fragmenting forces, which helped to disrupt the emergence of a common consciousness based on Shi'ism. For instance, the ethnic origins of the Shi'a populations were sometimes different. In Bahrain for example, the Shi'a population comprised indigenous Baharna Arabs and those of Persian extraction. The corporate nature of the communities was often disrupted along class lines. The importance of some members of the Shi'a in Bahrain and Kuwait as merchants and their consequent prosperity meant that they often had differing material interests from their poorer co-religionists. Sub-divisions within the Shi'a community on the basis of kinship were also important.

Secondly, one may look to the responses of the various states for the success in ensuring that the Shi'a disaffection that existed did not get out of hand. The response of the various governments in the Gulf states was twofold. On the one hand, there was an attempt further to appease the Shi'a communities. This took different forms in different countries. In Saudi Arabia, for instance, the governorship of the Eastern Province was taken away from the unpopular Ibn Jaluwi family. King Fahd's son, Muhammad, was appointed instead with a brief to be more receptive to the wishes of the predominantly Shi'a population. He was also invested with greater resources to distribute. On the other, the authorities attempted to reduce the potential potency of that threat. This often took the form of restrictive and coercive methods. The presence of the National Guard was boosted in the Eastern Province in Saudi Arabia during the Shi'a religious celebration of *ashura*. Intelligence and surveillance efforts were stepped up with great efficacy, especially in Bahrain. However, these methods were not always so negative. For instance, in Kuwait, as part of an apparent strategy to rehouse those living in the poorer districts of the capital, there was a hidden agenda of ensuring that the re-housing process lessened the concentration of Shi'a households. By 1987, all residential districts in Kuwait had less than 40 per cent Shi'a.

The experience of the 1980s has been to show that the internal threat to the Gulf states from its Shi'a population has been greatly exaggerated. Indeed, the political leaderships of the GCC states affected have shown themselves able enough to manage the threat which did exist. This was the situation which prevailed, even when there existed the external climate in which radical Shi'a opposition had the best opportunity to develop. Since the ceasefire in the Gulf, that external context has changed. There is no longer the possibility of a military breakthrough by Iran threatening the Sunni regimes of the Arab Gulf. Moreover, Tehran, with its new brand of pragmatic foreign policy, is committed to nurturing links with the Gulf states. With the external impetus to opposition now having been diluted, there is no question that this will deter potential recruits and demoralise and marginalise existing opposition groups. There can therefore be little doubt that the internal challenge from the Shi'a opposition will recede. Certainly, the potential for organised political challenges will fade, but the ability of existing groups to mount individual sabotage or assassination attacks should also diminish as their resource base grows more narrow.

The diminution of the Shi'a threat, both internally and externally manifested, should not be taken to mean that the stability of the Gulf states' regimes is guaranteed. Other areas of potential disaffection are often ignored when the challenge from one part of the domestic constituency dominates. Of course it often necessitates a catalyst to bring a challenge or at least the perception of a challenge to the surface. The radical Shi'a revolution in Iran was such a catalyst in relation to the Shi'a communities of the Gulf states, the potential threat from this quarter having been generally ignored prior to 1979. Predicting sources of instability is therefore problematic. Nevertheless, one can identify two areas which have the potential to provoke greater instability in the Arab Gulf over the next decade or so. These relate to the shrinking resource base relative to the demands being made upon it, and the problem of succession. In both cases Saudi Arabia is the main state to be potentially affected. However, it should be remembered that Saudi Arabia is the leading state of the GCC. Though it does not exercise a dominating influence over its fellow members, its fortunes nevertheless have a tremendous effect on them. Furthermore, Saudi Arabia is one of the three regional powers of the Gulf area. Any change in its stability and internal coherence will have a major effect on the power relations prevailing in the regional sub-system.

Oil is the fundamental component in the whole resource question. Up until 1982 most of the GCC states had established expenditure patterns which were predicated on the continued generation of high levels of oil income. Even those states relatively less well endowed with oil reserves, such as Bahrain, prospered as a result of the recycling of oil income from its more affluent neighbours. As the 1980s progressed, the oil market became profoundly depressed resulting in much lower prices. The Gulf area has been disproportionately affected in relation to the demand for oil. The net effect of this has been both greatly reduced exports of and revenues from oil, see Table 3.1. The GCC states have coped with this unforeseen problem in a number of ways. Many of the more capital-intensive infrastructural projects begun in the 1970s have been completed, resulting in lower levels of capital expenditure being made in any case. Some savings have also been made through reductions in government spending. Alone these cuts were insufficient. Increasingly, deficit financing was resorted to as a means to cushion the impact of the regional recession. The enormous accumulated resources of the main oil-producing Gulf states meant that this was a viable strategy in the short and medium term. As long as oil revenues were predicted to rise within a ten-year period the strategy remained a politically prudent one.

The central assumption of this deficit financing strategy, namely that oil revenue would recover by the early 1990s, is progressively being challenged by the 'new conventional wisdom' that rising real prices and a much greater call on OPEC oil will be delayed until the late 1990s or even the early part of the next century. If this proves to be correct, oil revenues will remain similarly depressed. The prospect for the GCC states is declining or stagnant growth, possibly into the next century. Even though there is evidence that the private sector, especially in Saudi Arabia, is playing a more mature role in the economy, it cannot be expected to take up anywhere near the whole slack caused by the retreat of the public sector. While certain Gulf states, such as Kuwait, are relatively well placed to ride the slump, not least due to their large reserves relative to the size of population, others, principally Saudi Arabia, will be faced with increasingly difficult choices.

One of the main ways in which Saudi Arabia was able to finance its budget deficit was in operating a large balance-of-payments current account deficit. The accumulated deficit from 1983 to 1986 inclusive was some $60 bn, see Table 3.2. The fortunes of the Saudi external economy was made good only by running down foreign reserves. Central Bank reserves, for instance, had fallen from near the $140 bn mark in 1982 to around $80 bn in 1986[15]. This in turn has compounded the revenue problem, as Saudi earnings from assets abroad have declined. In four years up to and including 1986 income from investment fell by an estimated $4.5 bn to stand at just over the $11 bn mark[16]. It would be rash to attempt to predict when such a strategy would no longer be viable. At present rates one can say that the Saudi external economy will be in serious straits just before the middle of the next decade. There are signs that the Saudi government is trying to act. Budgets have been progressively trimmed, while a precedent has been set for the culturally unpalatable course of borrowing by the flotation of a domestic bond issue. However, recourse to either massive cuts in budget spending or wholesale domestic or external borrowing at commercial rates would bring with it its own set of political and economic problems.

The resource challenge facing the Gulf states, and especially Saudi Arabia, is made far more acute by the demography of the area. At a time when the Gross Domestic Product of the GCC has been falling, birth rates have continued to rise unchecked at a high rate. Currently, the indigenous population of the GCC states is rising at between three and four per cent each year. The product of past increases is also being felt on the labour market, where the profile of female manpower, though starting from a low base, is rising. Moreover, those new arrivals on the labour market have grown up in the context of the conspicuous consumption which characterised the 1970s and early 1980s. Their expectations have been moulded by what was an exceptional period of growth in private and public consumption, rather than what may be described as the norm. The government will find it difficult to steer a course between rapidly growing unemployment and the temptation to expand state spending in order to provide jobs, which would in turn act as a force for increasing the budget deficit. The possibility of a growing crisis of expectation among the young also applies to those Saudis who have been educated to a high level, but who are unable to attain the positions of responsibility commensurate with their qualifications. In particular, they may come to resent the fact that historically senior government positions have been allocated on the basis of patronage rather than merit.

Growing dissatisfaction with the tightness of the jobs market, and the paucity of opportunities for promotion to senior positions could manifest itself politically. This could result in demands for greater political participation. Alternatively, it could push more and more of the younger, educated generations towards movements of protest such as the Sunni Islamist groups, one of which was responsible for the occupation of the Grand Mosque at Mecca in 1979. All such alienated elements could be expected to share a common unhappiness at the quality of the policies and management of the state. Were such a scenario to emerge, the political leadership in Saudi Arabia would be well advised to make reforms in good time and of their own choosing. Reforms should be aimed at increasing the circulation of elites, with merit playing a more important role. The formalisation of institutions giving greater access to political power and decision-making would also be important.

The second area of potential instability in the Arab Gulf could occur as a result of leadership successions. The question of succession is a particularly problematic one in the political culture of the Gulf states in any case. Primogeniture is certainly not the norm. The traditional bedouin notion of the succession falling to the member of the shaikhly family best qualified is still greatly admired in the society of the area. It is thus not regarded as illegitimate for a particular member of a family to advocate his own claims regardless of his seniority in terms of age or lineage. The nomination of heirs apparently cannot therefore be regarded as a guarantee of their consequent succession. The fact that the replacement of old and ailing rulers has taken place, often on a number of occasions, in the UAE, Oman and Qatar this century illustrates that the divine right of rulers does not prevail. Successions then are invariably times of flux, with a period of uncertainty following the demise of a ruler, even if the succession is broadly uncontested.

There are two states where successions at the highest level can be reasonably predicted to take place during the decade of the 1990s or soon after. These are in the UAE and Saudi Arabia. In the former the rulers of both the leading Emirates, Abu Dhabi and Dubai, are both very senior in years. Moreover, the ruler of Dubai, Shaikh Rashid bin Said al-Maktoum, has been chronically ill for some years. Both Shaikh Rashid and his counterpart in Abu Dhabi, Shaikh Zayid bin Sultan al-Nahiyan, dominate the politics of their Emirates and indeed of the federation. The council of ministers has been virtually inactive since the illness of Shaikh Rashid, its premier, prevented him from playing an active political role. Shaikh Zayid's authority as the president of the UAE is unchallenged.

Both men have chosen successors. In Abu Dhabi the heir apparent is Khalifa bin Zayid. In Dubai it is Shaikh Rashid's eldest son Maktoum. In neither case can these successions be assumed, although on the face of it they appear the most likely scenario. In the case of Abu Dhabi, doubts surround the leadership calibre of Khalifa. In the case of Dubai, Maktoum has a reputation for being relatively uninterested in affairs of state. The youngest of Shaikh Rashid's sons, the defence minister Shaikh Muhammad, is widely acknowledged as being more in the mould of his father. Assuming that both designated heirs succeed, they still face a challenging task in following their predecessors, especially in winning over the respect and loyalty of the royal and senior families and citizens of their own Emirates. Even if successful here, there still remains the question of the future of the federation. It was straightforward enough that the rulers of the largest most wealthy Emirates should lead the UAE as long as they were the eldest and most respected rulers in the federation. But what will happen when they are replaced by much younger and untried men? Will the rulers of the other Emirates be happy to see the leadership of the UAE go to the rulers of Abu Dhabi and Dubai *ex officio*? Or will they lobby for the introduction of a system of rotation in the presidency of the federation? Such questions of course cannot be answered. Suffice it to say that the demise of Shaikh Zayid and Shaikh Rashid could unleash additional forces for fragmentation within the federation.

If the future political leadership of the UAE is marginal to the fortunes of the GCC, the same cannot be said of Saudi Arabia. The poor health and consequently the age of the three most senior members of the royal family - King Fahd, Crown Prince Abdullah and Prince Sultan - mean that Saudi Arabia could be faced by not just one but as many as three successions over the relatively short period of the next decade. Were this to happen, the process would be extremely unsettling, even if the successions themselves took place without contention or opposition. The accession of a new monarch would no doubt be followed by personnel and policy changes as each man sought to stamp his own authority on the throne. There are other reasons why personnel changes would probably be made. Firstly, in order for the individual princes to reward those members of the royal family and senior commoners who have been their closest supporters and advisers. Secondly, because of the family connections of the men involved. While King Fahd and Prince Sultan share the same mother, Hassa bint Ahmad al-Sudairi, and hence are full brothers, Prince Abdullah has, in Bint Asi al-Shuraim of the Shammar tribe, a different mother. He would thus be expected to introduce more members of her tribe and from her region.

The rapid demise of the three most senior members of the royal family in Saudi Arabia would bring to an end the elite continuity at the very top. This could be especially pertinent in relation to the armed forces. Prince Sultan, the defence minister, has occupied this portfolio since 1962. He knows the workings of the armed forces intimately, and knows many of the officers personally. Similarly, Crown Prince Abdullah has been the head of the National Guard for some 26 years. It is more than likely that they would hear of dissent in the respective parts of the armed forces before it turned into serious grievances. Likewise, it is extremely unlikely that a group of officers could mount a coup d'etat without it coming to the attention of the respective princes. However, their replacement through death or as a result of accession to the throne would disrupt the close personal ties which exist between the highest levels of the royal family and the armed forces.

The argument has so far been advocated that successions taking place as a result of natural causes could be destabilising, even if there is broad consensus around the choice to be made of monarch. Such a smooth transition has characterised the last two successions: from King Faisal to King Khalid in 1975; and from King Khalid to King Fahd in 1982. Indeed, the Saudi royal family has shown itself to be adept at managing the succession question over the past four decades. However, the previous two successions took place in the context of economic well-being, during the height of the oil-related boom. It is more than likely that any successions taking place in the 1990s will do so against a backdrop of recession, retrenchment and increasing anxiety about the state of the economy. The tableau may be one of growing discontent from below, possibly resulting in increasingly different views among the senior princes as to the policy choices to be adopted in both the economic and political spheres. It cannot therefore be assumed that the consensus in the royal family will continue solely because disunity will jeopardise the position of the ruling dynasty. Indeed, arguments and dissensions may become more acute precisely because of these concerns.

As the economic context in which Saudi Arabia finds itself deteriorates, any divisions among the senior princes over the succession question are increasingly likely to be policy related. Personality differences and personal ambition may overlap, but would probably be secondary. Indeed, the routine speculation about factionalism in the royal family based on birth and personal jealousy is somewhat exaggerated in that it tends not to be a crucial factor in leadership successions. Of course, only one monarch has ever been unseated in Saudi Arabia, that being King Saud in 1964. This was not a function of factionalism or personal ambition. King Saud was manifestly inept, while his eventual successor, King Faisal, was extremely loath to set the precedent of deposing his elder half-brother.

It is, therefore, probably the case that speculation over whether, for instance, Crown Prince Abdullah will be happy with another member of the 'Sudairi Seven', namely Prince Sultan, succeeding him is largely idle observation. Whether he is happy or not is probably irrelevant. Sultan, as the incumbent second deputy prime minister - a post which effectively makes him the heir apparent's heir apparent -, will almost certainly in turn succeed. The most important decision will then be who will fill Sultan's shoes. The most likely candidate is Prince Nayif, the interior minister, who, at 56 years of age, is appreciably younger than his more senior kinsmen. He is also a full brother of Fahd and Sultan, as is Prince Salman, the governor of Riyadh, who is widely acknowledged as one of the more competent administrators within the upper echelons of the al-Saud family. Concern at the brothers turning the throne into the preserve of their wing of the family can be expected to mount. However, it is perhaps on the face of it unlikely that the 'Sudairi Seven' would be so precipitate as to attempt to formalise this tendency. This whole issue also raises the question of when the succession will finally drop a generation from the sons of the founder of the modern Saudi state, King Abdul Aziz, to his grandsons. As Abdul Aziz's youngest sons are actually younger than some of his eldest grandsons, the situation is also blurred in terms of seniority. With the former still being in their late forties, however, the problem of seniority among grandsons could well not present itself as a problem until the next century.

Although Dubai, Abu Dhabi and Saudi Arabia have been singled out for consideration in relation to the succession, there could well be problems faced in this regard elsewhere in the Arab Gulf. The situation in Sharjah appears still to be unsettled in the aftermath of the events which took place in 1987. This has proved to be the only entity in the Gulf to have experienced a major succession crisis during the life of the GCC. The situation in Oman may prove to be problematic if anything should happen to the present ruler, Sultan Qaboos. While his position is relatively unchallenged, there is no obvious successor, especially given the absence of a direct heir. Barring unforeseen eventualities, however, Oman is unlikely to be affected by a succession crisis in the 1990s.

A Framework for Gulf Security

So far the Geneva negotiating process has concentrated on the nature of the peace to be forged between Iran and Iraq. The painstakingly slow progress has been due to a lack of consensus between the two protagonists over the arrangements that should form the basis of the peace. In particular, the visions of Tehran and Baghdad have differed over the international boundary which is to prevail, especially along the Shatt al-Arab. Broader questions of state security and the security of supply routes are also at stake. The depth of mutual distrust and hostility has made the talks even harder to propel forward. Consequently, it appears that the peace talks will drag on for a long time, to be measured in months and even years. The process should eventually yield movement over certain issues, such as the repatriation of prisoners of war who want to return to their countries of origin. However, the eventual conclusion of a broad formal peace treaty between Iran and Iraq must be in doubt. The ceasefire that currently prevails could become institutionalised into a *de facto* arrangement governing bilateral relations between the two.

Whatever the final outcome of the UN-led peace process, it is clear that it will affect more than just the two central protagonists. In particular, the nature and depth of the arrangement which emerges between Iran and Iraq will be of vital interest to all the members of the GCC, even though they are not direct parties to the peace talks. The GCC states have a number of key interests, which necessitate that the organisation should play a leading role in the future of the area.

The first and most obvious interest is that a stable peace should be established. All the Gulf states were alarmed at the extent and rapidity of the metamorphosis of the land war into a Gulf conflict. The wider turmoil eventually embroiled them all indirectly to some extent; their security was threatened and their supply routes disrupted.

This chastening experience has dispelled all notions that it is in the interests of the smaller states of the area to have the regional powers distracted by war. As was shown in 1986 and 1987, the smaller Gulf states have no way of managing a conflict once it is in motion. Kuwait may have been successful in luring the United States into a military role in the waterway, but it was unable to influence the conduct of the superpower once it had become involved. Even Saudi Arabia, whose aspirations to being a regional power arguably gave it the most to gain from the Iran-Iraq war, came to appreciate that the negative aspects of the conflict were paramount. The war depressed the economies of the region, and was a major strain on the Saudi economy, which gave financial support to Iraq. The conflict also revealed that, despite Saudi Arabia's considerable military capability, it was vulnerable to attack. The attacks on its merchant shipping in the Gulf and, more particularly, the Mecca riot in 1987 well illustrated this point.

The second interest of the member states is that a stable balance of power should be established in the area. This would have two main effects. On the one hand, it would guard against one state dominating the region - a situation which would be unfavourable to the Gulf states as the smaller and weaker parties, and which in turn would inevitably reduce their parameters of action, and force them to appease the dominant power more intently. Moreover, it would encourage revanchist ambitions on the part of the other humbled regional power. The situation which prevailed in the Gulf in the 1970s was a good example of this phenomenon. Iran was capable of establishing its dominance over the region. This permitted it to seize the Tunbs islands and force an agreement over Abu Musa island on Sharjah, and to assume the dominant role as regional policeman. It was also able to foist a treaty on Iraq which forced Baghdad to make concessions over border sovereignty. Iraq resented this domination, and sought the circumstances, eventually provided in 1979, when it could attempt to correct the imbalance in the area.

On the other hand, it would be more likely that balanced power relations would result in a more stable peace, with both Iran and Iraq being deterred from initiating a further round of armed conflict. It could be persuasively argued that the reason why the war broke out in 1980 was as a result of the shift in the balance prior to that date. The apparent turmoil in Iran and the effect of the purges of the officer corps of its armed forces appeared crucially to weaken Iran. As a result of the seeming disappearance of the deterrent to war, Iraq took the risk of trying by force of arms to reassert its position in the area.

The third area of interest for the GCC states is in preserving the character o its own organisation. The GCC was formed at a time when both Iran and Ira were distracted by the war. The six would not therefore look with relish at th admission of one of those states, especially if it were the dominant power of th day and expected to stamp its interests on the GCC. There is certainly no rational for the member states to admit either of the regional powers. The smaller state have effectively prevented Saudi Arabia from dominating the affairs of th Council, and are therefore not in need of a second large power to balance th influence of Riyadh.

Some observers in the GCC states fear an attempt by Iraq to gain admission t the organisation. This would be most unlikely to happen in the short term, bu could take place after most of the peace negotiations are complete. If Iraq wa determined to join, it would be difficult to resist. In some respects Iraq is we qualified for membership, being an Arab state and having an outlet on the Gul The smaller Gulf states, especially those more vulnerable to Iraqi pressure, suc as Kuwait and Bahrain, would be unable to reject such an application. They woul turn to Saudi Arabia as the largest member to take such a difficult decisior Opinion differs on whether Riyadh would have the fortitude to stand out agains Iraq, in spite of the potential leverage which it possesses over oil matters. It i extremely doubtful that Riyadh would say no.

If an Iraqi application for admission looked likely the GCC might tak pre-emptive action. Rather than let Iraq gain control of what is a reasonabl coherent organisation, the GCC could admit other new members as a way o countervailing potential influence from Baghdad. This would also have the effec of dissipating the power of the Council, thus presenting Iraq with an empty shel rather than a vehicle for potential action. Potential new members to b approached would be the two Yemens, especially if they were to be united. Eve Jordan has been mentioned as a possible member. This course of action woul effectively emasculate the organisation which would probably cease to perforn any real function.

However, should Baghdad be admitted to an organisation that was small an wieldy enough to be an efficient instrument of foreign policy, then this coul precipitate resignations from the GCC. The state most likely to withdraw woul be Oman. Muscat is located far enough away from Iraq not to feel any direc threat, while Sultan Qaboos has deliberately kept a distance between his countr and the central issues of the Arab world. Notably, the emotional pull of Arat nationalism is far less in Oman than in even the other Gulf states, thus making it secession less momentous. Whether or not Oman would resign if Iraq were to joi the GCC, would probably depend upon the nature of the peace. If Iraq had foiste an inequitable peace upon Tehran, for instance by taking sovereignty over th whole of the Shatt, then Oman would feel particularly wary about the future. I such a situation Muscat might feel that the seeds of the next war had been sow ıd that the admission of Iraq to the GCC had ensured that the next war would ; between Iran and an enlarged GCC, rather than between Iran and Iraq.

Some of the GCC states obviously believe that they have a role to play in trying recreate a balance of power in the Gulf. In early August 1988, just after Iran ccepted Resolution 598 the Omani Minister of State for Foreign Affairs toured idely around Iran during a one-week visit. He returned for a second visit in early eptember. Though Oman has emerged as the chief interlocutor between the CC and Iran, other states have been keen to encourage the new moderation in ehran. King Fahd delivered an extremely conciliatory speech in the aftermath f the 1988 pilgrimage. Kuwait resumed diplomatic relations with Iran in October 988. A month later Gulf Air, the carrier part owned by Bahrain, the UAE, Qatar nd Oman began making arrangements for a resumption of flights to Iran[17]. hese moves show quite plainly that the GCC states are adopting a conciliatory ance towards the Islamic republic.

Most of the moves made so far are aimed at ensuring that Iran, from a omentary position of weakness, is not alienated from the Gulf sub-system of ates. However, there are indications that the GCC states are giving thought to uestions of longer-term stability in the Gulf. The idea of some sort of permanent amework could emerge as a way of institutionalising a new arrangement for Gulf curity. The idea has also been taken up by non-Gulf Arabs. The Soviet Union also believed to be interested in the concept and to be giving support.

The new framework would be based on three principles which all the littoral ates of the region would be expected to share. They are that: all have a common terest in the Gulf war not being restarted, all have a desire for stability and curity in the Gulf, and all have a role to play in ensuring the continuation of this curity. Although there might be varying visions on how this would be organised, s essential characteristic would be that it would include Iran as well as Iraq and ıe GCC. The latter organisation would have a pivotal role to play akin to that of n honest broker in promoting trust between the former antagonists.

Two arguments are put forward for establishing a permanent structure. First, ıat a formal structure is needed if the disputes which will inevitably arise from me to time among the various members are to be nipped in the bud or at least ıanaged effectively. Secondly, that a permanent forum for regular meetings and onsultations will help promote trust and understanding among all the member tates thereby reducing the possibility of costly miscalculations. This role would e facilitated if the framework were to be given a wider brief to include, for ıstance, a role in the economic affairs of the area.

It is still premature to talk about what form such a framework would take and ow wide its functions would be. It is likely that it would be more broad-based ıan simply concerned with security affairs. Although Gulf security would be its principal motive, like the GCC itself the structure would be given other functions. Chief amongst these would probably be an economic function directed at the growth and reconstruction of the area. The structure could also act as a conduit for finance aimed at helping to fund the reconstruction process. Such an enlargement of the framework would instil greater moral impetus into the whole scheme.

Conclusion

The GCC has acquired a generally negative reputation. While the organisation certainly has its shortcomings, this does not tell the whole story. The individual members of the GCC regard the organisation more positively, and hence the end of the Gulf conflict has not thrown into doubt its continuation in the future. In particular, the existence of the GCC has enabled the Arab Gulf states to adopt united stands on a range of issues, with the smaller and more vulnerable members in particular given confidence by the relative anonymity rendered by the Council. The greatest fear on the part of the smaller members of the six, namely domination by Saudi Arabia, has not so far come to pass. That said, there are a number of intrinsic reasons why the GCC will never be a close-knit and cohesive multilateral unit. These factors include: the differing geopolitical priorities of its members; anxiety about domination in a more closely co-ordinated unit; and traditional and continuing frictions amongst its constituents.

With the Iran-Iraq war apparently at an end, the major challenge to the regional sub-system is the re-creation of a balance of power in the Gulf area. The GCC has both strong interests in this respect and the ability to play an important role in the establishment of a stable and lasting equilibrium. The two greatest challenges are: to ensure that Iraq does not use the short-term advantage it has obtained from the war to achieve a temporary dominance in the area; and, as a consequence of that, to ensure that a *casus belli* is not created which makes it inevitable that Iran will seek recourse to war to correct perceived injustices perpetrated in the aftermath of the Gulf conflict. The best way for lasting security to be established is through the creation of a new multilateral framework embracing the whole of the Gulf, and to consist of Iran, Iraq and the GCC as equal partners. This 'super GCC' would be best placed to manage disagreements when they emerged and would help build understanding and trust between Tehran and Baghdad.

Notes

[1] For example the foreign ministers meeting which took place in Muscat in 1976.

[2] Economist Intelligence Unit Regional Review for the Middle East and North Africa, 1984, p.24.

[3] Economist Intelligence Unit Regional Review for the Middle East and North Africa, 1986, p.16.

[4] Interview with Shaikh Salim, then Minister of Defence of Kuwait, 3 December 1987.

[5] Laura Guazzone, 'Gulf Cooperation Council: The Security Policies', *Survival*, March/April, 1988.

[6] Interview with Shaikh Salim, op. cit.

[7] *The Times,* 12 January 1988.

[8] It should be noted that the US ambassadors in many of the Gulf states were political appointees during the second Reagan presidency. These included the heads of mission in Manama, Doha and Muscat.

[9] The distance from Kuwait City to Muscat is some 750 miles.

[10] Though the whole of the Strait of Hormuz falls within the territorial waters of Iran and Oman, it is classified as an international waterway, and hence international shipping have rights of passage. There were instances when the US navy patrolling in the area showed less than a full understanding of this situation. On occasion US naval ships warned Iranian craft against attacking shipping because they were in 'international waters'. The Iranian navy then made things difficult for the Omanis by quoting the Americans when challenged over operating in what were in fact Omani waters.

[11] W. E. Moubarak, 'Kuwait's Quest for Security: 1961-1973.' Unpublished PhD thesis, Univ. of Indiana, 1979, p.82.

[12] *Mideast Market*s, 21 July 1986.

[13] One prominent member of the Bahraini royal family wanted to launch a military attack against the Qatari force.

[14] *Mideast Markets*, 25 July 1988.

[15] Reserves for the GCC states ($ million) for 1986.

S. Arabia	Kuwait	UAE	Qatar	Oman	Bahrain	GCC
80,000	86,000	22,000	12,000	3,100	1,600	204,700

Estimated totals to end of period.
The National Bank of Kuwait, *Gulf Cooperation Council in Figures*, 1988/89, p.33.

[16] Investment income for Saudi Arabia ($ million).

	1982	1983	1984	1985	1986(p)
Investment income	14,059	15,867	13,365	12,418	11,279

(p) Provisional
The National Bank of Kuwait, *Gulf Cooperation Council in Figures*, 1988/89, p.16.

[17] BBC/SWB/ME, 24 November 1988.

Table 3.1 Net Oil Export Earnings of GCC States 1983 - 1988 ($ billion)

	1983	1984	1985	1986	1987	1988
Bahrain	0.22	0.18	0.17	0.08	-	-
Kuwait	7.91	8.50	7.86	5.28	6.0	5.6
Oman	4.04	4.09	4.76	2.49	-	-
Qatar	3.29	4.29	3.07	1.74	1.9	1.5
Saudi Arabia	47.22	40.41	27.55	19.55	21.4	19.9
UAE	12.02	13.72	12.27	7.27	8.8	6.5
Total	74.7	71.19	55.68	36.41	-	-

Source: Secretariat estimates, *Energy in non-OECD Countries*, OECD/IEA, Paris, 1988. 1987/88 figures from *Petroleum & Energy Intelligence Weekly,* November 14th, 1988.

Table 3.2 Balance of Payment Current Accounts of GCC States ($ million)

Period	S. Arabia	UAE	Kuwait	Qatar	Oman	Bahrain	GCC
1982	7,571	4,877	7,001	1,127	460	172	21,209
1983	(16,851)	5,287	5,257	410	359	123	(5,414)
1984	(18,402)	6,375	7,464	830	107	18	(3,608)
1985	(12,936)	4,816	6,946	549	110	611	97
1986	(11,936)	5,392	2,370	(189)	(996)	95	(5,264)

( ) figures in brackets indicate negative balances.
Source: *Gulf Cooperation Council in Figures 1988/89*, The National Bank of Kuwait, Kuwait, 1988.

4. THE SUPERPOWERS AND THE GULF

Introduction

The establishment of a ceasefire in the Iran-Iraq war has reduced the tension in the Gulf waterway and foreign powers have begun to scale down their naval fleets in the area. Nevertheless, the cessation of hostilities has not removed the sense of amazement at the events which took place in the Gulf during 1987 and the first four months of 1988. The US in particular concentrated a staggeringly large naval force which in November 1987 totalled 17 vessels in the waterway, with a 15-ship flotilla hovering just outside. Altogether it was the largest fleet put together by the US since the Vietnam war. The other powers also raised their naval profile. The French increased their force from three to ten while Britain's Armilla Patrol was boosted from three to seven ships. The Soviet presence rose from two ships in the Arabian Sea to six in the Gulf. Even three of the smaller European states, which are not now noted for adventurism far from their home shores - Belgium, Holland and Italy - had minesweepers operating in Gulf waters. Finally, and by no means least remarkably, the normally reticent Japanese seriously toyed with the idea of sending coastguard vessels to the area[1]. Instead, they undertook to fund the placement of a navigation system along the western shore of the Gulf.

The sudden and rapid growth in foreign, and especially Western, forces in the Gulf was unpredicted, even a matter of weeks before it occurred[2]. Indeed, the whole involvement caught public opinion in the West unawares. Yet the fact that the massive build up of forces took place says much about the West's perception of its interests in the area. The commitment of forces on such a scale, with the financial costs and risks to military personnel involved, suggests that the West in general and the US in particular felt that its vital interests were in jeopardy. The fact that the press and the general public alike were unprepared for this massive increase in forces strongly implies that most people remain generally ignorant about their country's national interests in the Gulf region. In the light of this, any discussion of the superpowers and the Gulf must be preceded by a brief survey of the national interests of the two superpowers.

The United States

In justifying the increase in naval forces in the area, many references were made to US strategic interests in the Gulf. Yet somewhat surprisingly few attempts at the time were made publicly to state what those interests were. Moreover, the naval build-up was justified in terms of much rhetoric about freedom of navigation, while it was plain that the navy's brief did not run to such a remit. With so many false references and erroneous justifications clouding the issue, it is important to start this section by going back to basics. What are the US national interests in the area and what problems do they present?

It is ironic, in view of the general ignorance on the matter, that three of the core US interests in the area have existed for some four decades. They are:

Maintaining the unimpeded flow of oil from the Gulf to the West. This is based on the premise that oil is the 'lifeblood'[3] of the economies of the industrialised nations as well as of many states in the developing world. Any significant disruption in Gulf oil supplies, the argument goes, would cause world oil prices to skyrocket, plunging the market economies of the free world into a crisis akin to those experienced in 1973-4 and 1979-80. Counter-arguments criticising the continued US preoccupation with the oil issue, based on the fact that only 5 per cent of US domestic consumption currently comes from the Gulf, are dismissed. The fact that Western Europe receives 30 per cent of its oil from the Gulf and Japan almost 60 per cent is held up as amounting to Western strategic dependence. It is pointed out that if the supply to Europe and Japan was cut off, the US would certainly suffer indirectly through the effect on prices that the sharp reduction in supply would engender. It is further noted that the two previous oil price shocks were triggered by less than a 5 per cent reduction in supply. In 1987, the Gulf accounted for 22 per cent of the world's oil production.

Blocking Soviet expansion southwards. The US seeks to stymie attempts by Moscow 'either to control directly or to increase significantly its presence or influence over the region'[4]. This view is predicated on the basis that the Soviet Union has long-term designs on the Gulf, regardless of the 'new thinking' taking place in the Kremlin. Furthermore, the US claims a special interest in the area, as compared with Moscow, because oil imports are relatively unimportant to the

Soviet bloc, given the USSR's status as a net exporter. While this objective owes its origins to the 'Cold War' and is a product of an essentially bipolar view of the world, it was given added impetus by the Soviet invasion of Afghanistan in 1979. The direct role which the US claimed for itself in the context of this aim was enshrined in the 'Carter Doctrine' of 1980. Under the terms of this policy statement, President Carter pledged that: 'Any attempt by any outside force to gain control of the Persian Gulf region will be regarded as an assault on the vital interests of the United States of America'. The pledge went on to state that the US was prepared to back up its interests with action, saying that 'such an assault will be repelled by any means necessary including military force'[5].

Maintaining an active presence in a geostrategic part of the world. The Gulf area occupies a vital location between three continents. This broadly explains the institutionalisation of the US naval presence in the Persian Gulf and the Arabian Sea which came about with the establishment of the Middle East Force in 1949. An important adjunct of this policy is to establish and maintain open lines of direct communication with the states of the region through the medium of diplomatic relations. To that end, the State Department's underlying view is that the absence of diplomatic links with Iran is an aberration which should be as short-lived as possible.

Such a triadic prescription was straightforward enough until the revolution in Iran in 1979. Once the new Iranian regime had unequivocally, and speedily, made its attitude towards the United States plain, a further strategic imperative was added to the checklist to guide US foreign policy:

The confining of the Islamic revolutionary model of government to Iran. This involves the prevention of both its export and emulation in neighbouring states. Iran's policy after the revolution and during the war with Iraq was viewed as one of expansionism, which the US labelled a 'special danger'[6]. The ending of the war will not in itself change this attitude. US officials emphasise that many of the aims of the Islamic regime in Tehran are inimical to the West. Many of them have not been disowned, in spite of the ceasefire. The most obvious, it is said, is the open hostility towards the US. Iranian attempts to create instability in the moderate Arab states of the Gulf and attitudes towards energy policy, notably the desire to raise the price of oil, are also mentioned.

The establishment of the ceasefire in the Gulf has not substantively changed US interests in the region. The long-term preoccupations of safeguarding oil supplies and vigilance against potential Soviet attempts at subversion will continue. It also remains out of the question that the relationship between the US and post-revolutionary Iran will ever return to the quality of that experienced between Washington and Tehran during the days of the Shah. Indeed, the geographical importance of the Gulf continues to be valid. However, an end to the war and the changes brought about as a result of the war could well shift the relative importance of these aims.

The USSR

No discussion of Soviet foreign policy can omit reference to the profound changes in strategic thinking which have been taking place in the Kremlin since the mid-1980s. The advent of the Gorbachev years has brought almost every aspect of Soviet policy under the influence of *perestroika,* and foreign affairs is no exception. The Soviet Union itself refers to the changes which have taken place in this sphere as being 'new thinking'. For the purposes of a study focusing on the Gulf, five pillars of this new thinking can be identified which are a departure in substance or emphasis from earlier Soviet foreign policy. They are:

A preoccupation with domestic policies, which in turn devalues the importance of foreign affairs. There is a realisation that the reform of the Soviet Union domestically, both politically and economically, must take precedence over less pressing goals in the foreign domain. The most important manifestation of this view is that fewer resources will be devoted to the pursuit of foreign policy aims in the future. As one former British ambassador to Moscow has written: 'the Soviet Union requires that foreign policy should not obstruct the implementation of internal reform and should if possible facilitate it'[7]. Already this has shown itself in the Soviet reluctance to extend military credit to its traditional allies, and the demand that repayments should be made promptly and in hard currencies.

The 'de-ideologicalisation' of foreign affairs. The Soviet Union will no longer take sides in disputes simply because one of the protagonists is identified as having a 'progressive' or 'socialist' label. Indeed, the notion of promoting friction between states in order to increase Soviet influence appears to have been dropped completely. Rather it is the threat to peace and stability which now governs Soviet reactions to regional conflicts, rather than the ideological complexion of the states involved.

The recognition that other states have legitimate, vital interests. This appreciation certainly extends to the US as a superpower. One can only speculate as to what would have been the Soviet reaction to the massive US naval build-up in the Gulf had it taken place even five years earlier. It is likely that the Soviet Union would have been far less accommodating than it was in 1987. However, this greater sensitivity to the core issues of other international actors also extends to smaller regional states.

The belief that regional conflicts must be ended through the forging of political solutions. The Soviet Union is increasingly anxious not to get embroiled in military conflicts as a method of trying to solve problems. This includes direct involvement, such as over the Soviet military presence in Afghanistan. It also extends to indirect involvement through proxies, as with the Cuban presence in Angola, and through the continuous arming of Soviet allies. The refusal to give Syria an open-ended commitment to supply large quantities of the most up-to-date military equipment in order that Damascus may pursue its dream of strategic parity with Israel is one striking example.

The de-linking of regional conflicts from strategic questions of superpower relations. This principle reflects the fact that Moscow has ranked those foreign policy areas which are most vital to it, the most important being relations with the US and arms control. The thinking in the Kremlin takes the view that while some regional questions are important, they do not merit jeopardising more central issues. It now becomes possible, at least from the Soviet point of view, to have cordial ties between Moscow and Washington simultaneously with areas of superpower competition. A good example of this is Afghanistan, where the US continues to aid the opposition guerrillas as Soviet troops pull out. Though the guerrillas have continued to attack the Soviet forces in the country, and are steadily undermining the precarious regime left in Kabul, Moscow has not allowed this to interfere with the development of an understanding with the US.

These are some of the new guidelines which senior Soviet officials and commentators say will govern their policy formulation in the future, but there is no guarantee that they will indeed rule every area of Soviet activity. Furthermore, the Gorbachev leadership in Moscow has come to appreciate the difficulties in introducing a new philosophy to Soviet policy-making, both at home and abroad. Many areas of the bureaucracy have been loath to co-operate with the reforms, and there is no reason to think that foreign and defence matters will be any different from internal restructuring. In addition, there is always the worry that Gorbachev himself will be forced to abandon the substance of his programme, although it looks safe at least until 1990.

Nevertheless, a certain degree of caution in the West is both understandable and proper. Even if the thrust of the new thinking is implemented, it is too early to know exactly how this will manifest itself in a broad Middle East context. Apart from a general warming in relations with Israel and Egypt, and the appearance of less tolerance towards Syria[8], a clear strategy has not yet emerged. However, even after only three years in power, experience shows that there have indeed been substantive changes in the operation of Soviet foreign policy. In the Middle East context the decision to withdraw from Afghanistan is the major example.

Perhaps the last word on the new thinking which is under way in Moscow comes from a Soviet foreign policy specialist in North America. His conclusion on the changes are that it is very difficult for Sovietologists today to rule out almost anything with respect to changes in the Kremlin's thinking[9].

Having set the scene for the new Soviet thinking on foreign policy, we may now proceed to a consideration of Soviet interests and policies pursued with respect to the Gulf. The Soviet Union's core strategic interests in the area may be listed as follows:

The security of the USSR's southern borders. The first priority of the Soviet state must be the security of its territory and, in the context of this study, its boundaries with the Gulf area. Security does not simply extend to defending its borders against physical incursion. It also includes, for instance, blocking the importation of material which could be perceived as likely to subvert the state. This latter consideration is particularly pertinent in the southern Muslim republics of the Soviet Union, especially where large numbers of Shi'a Muslims are found, notably in Azarbaijan. Indeed, there are definite links between the Azaris of the southern Soviet Union and their fellow people in northern Iran. Sensitive to the demographic shifts taking place in the Soviet Union in favour of its Muslim peoples, Moscow is vigilant to ensure that the Iranian paradigm of revolution does not spread across its borders.

The increase of Soviet influence in the region. In the past the Soviet Union has virtually been shut out of the Gulf region. Moscow has enjoyed good, but not warm relations with Iraq since 1968. However, this has had negative consequences for its relations with other states in the area, as Baghdad's negative image among the lower Gulf states has helped to damage the USSR's cause. Until relatively recently among the other Gulf littoral states, Moscow enjoyed diplomatic relations only with Kuwait. And even then, Kuwait did not countenance internal communist activity. In the 1970s Iran and Saudi Arabia formed the basis of the United States' 'twin pillars' policy in the region, whereby trusted proxies had *de facto* responsibility for the security of the Gulf. Since the outbreak of the Iran-Iraq war, Moscow has managed to establish formal ties with the UAE, Oman, and most recently Qatar, although its cause had earlier been damaged by the invasion of Afghanistan. The highly reasonable hope in Moscow is that the withdrawal of Soviet troops from Afghanistan will remove the last impediment to the exchange of ambassadors with Saudi Arabia, and the acceptability of Soviet influence in the Gulf area.

The avoidance of confrontation with the West. Though the Soviet Union would dearly like to expand its influence in the area, it is not prepared to jeopardise central foreign policy objectives, such as superpower detente, in order to do so. Certainly, Moscow has eschewed the opportunity of scoring short-term propaganda points in the region at the expense of the United States. Its restraint in not seeking to exploit the shooting down of the Iranian airliner in July 1988 in the same way as the US did when Soviet jets downed a Korean civilian aircraft in 1983 illustrates this point.

The weakening of Western influence in the area. Given that the improvement in bilateral superpower relations has still not spread to regional issues, this objective remains part of the Soviet political agenda. Nevertheless, it is important to note that the emphasis given to it is far weaker in comparison with the orthodox approach of the Brezhnev years. Thus, it was over a Gulf issue, namely the forging of Security Council Resolution 598, that all permanent members of the United Nations Security Council co-operated in drafting their first unanimous mandatory resolution. This showed the Soviet Union willing to give its support to such multilateral initiatives, even if they originated in the West, providing that Moscow believed that they were in its best interests. Western sponsorship was not in itself an obstacle. Ultimately, however, the Soviet Union had second thoughts when the passing of the resolution coincided with a further increase in the size of the US fleet in the area. The Soviet authorities suddenly grew anxious lest they were simply legitimising policies which were solely in the West's interests. This anxiety was compounded by a concern at the possible spoiling role which Tehran might play over the Soviet withdrawal from Afghanistan. Consequently, the USSR shrank from endorsing a follow-up resolution calling for an arms embargo against Iran.

The USSR and Iran

The US failure to establish relations with Iran on a sound and stable basis has given rise to concern in the West that the way might be left open for the Soviet Union to form a strategic relationship with Tehran. This anxiety became especially acute in 1987 when, in addition to the US, bilateral relations between both Britain and France and the Islamic republic deteriorated sharply. As such, this fear was more a function of the decline in relations between the US and some European states on the one hand and Iran on the other; it was not necessarily predicated on any qualitative improvement in Soviet-Iranian relations. In seeking evidence of an emerging Soviet-Iranian bond, there has been a tendency to focus on the number of formal meetings between Iranian and Soviet representatives rather than looking to see if any positive outcomes have emerged from these talks. In other words, there has tended to be a preoccupation with the form of such contacts without attention to the substance or lack of it. There is also a tendency to forget the relationship which existed between the Soviet Union and Iran under the Shah. Though Iran was firmly orientated towards the West, it could not ignore the superpower on its borders. In fact, the 1960s and 1970s were marked by a high level of trade between the two sides[10]. The relationship, although politically frosty, was not allowed to deteriorate into hostility, in part at least by the maintenance of commercial channels between the two states.

Given the proximity of the Soviet Union and Iran and their aspirations to play an active diplomatic role in the area, there is a latent potential for suspicion and competition. The long historical legacy of bilateral conflict and tension would dispel any doubts on this point. The nineteenth century was full of instances of Russian interference in Persian domestic politics and rivalry with the British over influence and economic benefits. In the present century, there have been two occasions when Iranian territory has been occupied by armies from the north, one involving Russia in 1908 and the second the Soviet Union in 1941. The most alarming conclusion from an Iranian point of view was that the incursions took place at a time when regimes of very different ideological complexions were ensconsed to the north. The two examples thus illustrate a continuity of historical experience. They demonstrate that first the Russian Empire and then the Soviet Union have shown a recurring geostrategic interest in Iran. The actual experience of the occupations also illustrates the enormous influence that Iran's towering neighbour to the north can exercise over its internal affairs, not least with reference to the minorities issue.

Most recently, Iran was shocked by the Soviet invasion of Afghanistan, coming at a time when Tehran was in the midst of revolutionary turmoil. The deep distrust pervading Iran's view of the Soviet Union is underscored in a revolutionary context by two of the central tenets of the ideology of the Islamic regime. Firstly, the Soviet Union is an atheistic country. This denial of the existence of God makes it difficult for a regime which regards its faith as the origin of its liberation to pursue cordial bilateral ties. Secondly, the main secular principle of Islamists and nationalists alike in Iran is that the country should exist and function free from foreign intervention. The modern history of Iran is a catalogue of national humiliation born of the excessive interference of foreign powers in its domestic affairs. Having thrown off the US domination during the semi-independence experienced in the post-Mussadegh era of the Shah's rule, Iran will not easily exchange the Soviets for the Americans.

The Soviet Union welcomed the revolution in Iran in 1979. But this reaction was based more on relief at the downfall of a regime which was firmly orientated towards the US, rather than any positive regard for the Islamic system of government that emerged soon after the turmoil subsided. Within a year of the overthrow of the Shah the new administration in Tehran was reminded of the enduring threat of the Soviet Union by the invasion of Afghanistan. In retrospect it is no surprise that the pro-Moscow communist party, the Tudeh, was perceived by the new rulers in Iran as a fifth column. It was suppressed in the spring of 1983, at a time which also saw the expulsion of 18 Soviet diplomats and the launching of a campaign on behalf of the 'oppressed Muslims in the USSR'. This stark reversal in attitude by the Iranians stripped away any pretence of the existence of cordial ties; it also delivered an ominous threat to the national interest of the Soviet Union by threatening to stir up its southern Muslim peoples.

Since then Moscow has tried to use formal visits and economic ties to ensure that bilateral relations remain civil, but has done so without compromising its political line in the area. Former President Andrei Gromyko was never afraid to speak candidly when meeting visiting Iranian officials. He chided the Iranian Foreign Minister for the continuation of the Iran-Iraq war[11], and generally issued warnings to Iran holding it responsible for the activities of Afghan guerrillas on its soil[12]. Soviet officials visiting Iran have also made the point that relations with Tehran are 'much cooler' than the 'warm friendly atmosphere' prevalent when on a trip to Baghdad[13]. It is true that the Iranians have sidled up to the Soviet Union. However, this has been for tactical reasons. The Afghanistan issue has continued to be one over which they have refused to make substantive concessions. Their support for the Afghan opposition groups based in Iran and their desire to see the Soviet Union withdraw has been an unbending objective.

The one area in which the Soviet Union has attempted to woo Iran, and with disastrous consequences, was in the aftermath of the adoption of SCR 598. Having had second thoughts about the wisdom of endorsing the Western-led plan to isolate Iran, the USSR decided to pursue a unilateral path aimed at persuading Iran to co-operate with the UN. In this, Moscow was no doubt hoping to exploit its diplomatic lines of communication with Tehran, with the ambition of playing a vital role in the forging of a settlement to the conflict in the area. Allied to this move, the Soviet Union also called for the replacement of the individual naval presence in the Gulf by a UN force. For four months the Iranians gave Moscow enough encouragement to continue this line of action, but without ever themselves intending to comply with SCR 598. The Iranian reaction towards the Soviet moves had therefore been purely tactical. In December 1987, Moscow admitted the fruitlessness of its actions, saying that it had been 'too optimistic' with regard to Iran's willingness to co-operate with the UN. More profoundly, this whole chapter had betrayed an important error of judgement on the Soviet side as to the motives and intentions of Iran within the war. Moscow had hoped that its relations with both combatants would enable it to mediate between them. Ultimately, this proved to be fanciful and naive. Moscow was blamed for the failure of the UN initiative by the US and Britain[14], a call which the Arab Gulf states, including Iraq, quickly took up[15]. In return for four months of considerable diplomatic discomfort, the USSR had little to show for its efforts.

Both sides perceive that it is in neither's interests to allow their relations to deteriorate. The USSR does not want to provoke Iran gratuitously into trying to play a spoiling game over its scheduled withdrawal from Afghanistan. Moreover, it does not want the spectre of Soviet interference in Iranian affairs once more to be raised since it fears that in such circumstances Iran would again be driven to seek a close relationship with the US to counterbalance the potential Soviet threat. Yet neither does it really believe that the low level of relations between Tehran and Washington will continue in perpetuity. Sooner or later, the conditions will exist which will allow the two sides to establish correct diplomatic relations. The USSR knows it cannot prevent that. It simply seeks to ensure that both superpowers enjoy similarly correct relations with Iran, rather than seeing itself politically excluded as happened under the Shah. Though this may be the aim of foreign policy makers in the Kremlin, there is no guarantee that it can be attained. In particular, it is dubious if the Soviet Union, because of its size, location and historical relationship with Iran, can prevent itself from being enlisted in the internal political machinations of the Islamic republic over the next decade.

The USSR and Iraq

The relationship between the USSR and Iraq has had its ups and downs over the last two decades. Yet if one considers the various upheavals which have taken place in the region and, more recently, in Soviet policy towards the Middle East, there has been a remarkable degree of continuity in the quality of the ties. Nowhere has this been more evident than in the field of arms supplies, which has been the most obvious manifestation of the enduring links between the two states. Over the last 20 years the Soviet Union has been Iraq's major arms supplier. At certain times it has reduced these supplies when it has been dissatisfied with certain of Iraq's actions, such as its original decision to move its forces across the border into Iran. Owing to these uncertainties Iraq has, as noted earlier, diversified its arms procurement policy, notably in the direction of France. But, despite the occasional deterioration in the links, the Soviet-Iraq relationship has persisted. Indeed, throughout the war Moscow continued to be Iraq's premier arms supplier. In the latter stages of the conflict, when concern rose about Iraq's ability to continue to withstand the Iranian onslaught, Moscow increased the quality of its arms transfers, to include the most advanced aircraft, the Mig 29.

The relationship between the two states is founded firmly on mutually perceived utility, tinged in the past by the attractiveness of the Soviet economic and political model to the Ba'thist leadership. It is not the product of a patron-client relationship. Indeed, Moscow was powerless to influence Iraqi policy over the start of the war. Nor is it based upon any real cordiality, born of personal warmth or trust between the respective leaderships. Even the ideological affinity between the two regimes should not be exaggerated. Although at varying times Iraq and the Soviet Union have found broadly common ground in relation to certain issues, notably the Arab-Israeli conflict following the 1967 war and the importance of a centrally planned economy, this has not been the product of a common ideology. Ba'thism, which is essentially a form of pure Arab nationalism, is by no means compatible with the fundamental precepts of Marxism-Leninism. The Ba'thist leadership has tended to draw on the Soviet experience, especially in relation to the organisation of the economy, because of the appeal of tight centralisation to a party whose formative experience was spent underground hatching clandestine conspiracies. In foreign policy terms, all that has tended to exist has been an overlap in the pursuit of policies which may be somewhat vaguely described as 'radical'. The new thinking of the Kremlin on foreign policy, together with the new moderation in Iraqi foreign policy, means that while this radical overlap is likely to wane in the future it may well be replaced by a considerable degree of moderate overlap.

The existence of a strong and vigorous communist party in Iraq has complicated the ties between the two states. In the early days of the relationship the presence of the Iraqi Communist Party (ICP) looked set to deepen bilateral ties. The expansion of the Iraqi government into a Progressive National Front, which included the ICP, in 1973, helped the bilateral relationship. The signing of a Soviet-Iraqi Friendship Treaty in 1972 illustrates the extent to which the early 1970s were the apex in relations. However, the inter-party alliance was only a relatively short-lived affair, born of a tactical motivation on the side of the Ba'th, which wanted to settle the Kurdish question. By 1978, with the Kurdish opposition groups quiescent, a complete break between the two parties had taken place and the ICP had been driven underground. Suddenly the ICP became a barrier to warmer Soviet-Iraqi relations. The purge of the ICP certainly put a great strain on them, but Moscow did not end its relationship with Iraq because of the ICP. Nevertheless, even though Moscow abandoned the ICP to its fate, the break between the Ba'th and the communists did impair inter-state relations in so far as it retarded inter-party relations. The existence of a strong and alienated Iraqi communist party has prevented the Communist Party of the Soviet Union (CPSU) from developing the type of inter-party relations that it has forged with the Syrian Ba'th Party.

Given the lack of roots upon which the bilateral relationship is founded, it is perhaps surprising that this mutual utility has continued for so long. A number of factors have contributed to the longevity of the relationship. Firstly, there has on the whole been an absence of core disruptive issues. The Soviet Union was not, for instance, willing to allow the persecution of the ICP in the late 1970s to obstruct valuable inter-state ties with Baghdad. Soviet distress over the Iraqi invasion of Iran was a rare example of such a disruptive issue, however. Secondly, the political geography of the area has helped to lessen the potential strains between the two countries. The absence of a common border means that the tension which has existed between Iran as a regional power and the Russian Empire, and later the Soviet Union, has not been replicated in Iraqi-Soviet relations. The presence of Turkey as a member of NATO and an Iran where, under the monarchy, a strategic relationship existed with the US also increased the attraction of Iraq and the Soviet Union to one another. Finally, leading on from this point, the utilitarian nature of the relationship has been enhanced by default. The absence of alternative regional and external allies has meant that Moscow and Baghdad have tended to drift back towards each other, even after the occasional disruption in relations. In the USSR's case, Turkey and Iran - at different times for different reasons - were beyond cultivation, as has been Saudi Arabia. For Iraq, the implacable enmity of the Zionist lobby in Washington has greatly restricted the potential of relations with the US.

By the time the Gulf war came to an end the quality of relations between Iraq and the Soviet Union was reservedly good. The Soviet Union, which has always been committed to the continued integrity of the Iraqi state, backed this up with massive weapons transfers during the last two years of the conflict. However, Baghdad was extremely suspicious of Moscow's overtures to Tehran in the wake of the Security Council's adoption of Resolution 598. If Iraqi suspicions were heightened in late 1987, they dissipated as the Soviet Union failed to exploit its line of communication to Iran. By July 1988 the Soviet First Deputy Foreign Minister, Yuli Vorontsov, after a visit to Baghdad, spoke of the 'warm, friendly atmosphere' in which the talks had taken place. A senior member of the CPSU dealing with international affairs summed up the state of the bilateral relationship the following autumn. Speaking in October 1988 he pronounced the relationship as being 'not at its peak, but not on its way down'. The future relationship between the two countries is almost certain to continue to be founded on the basis of mutual utility. This is certainly the drift of Moscow's foreign policy in the region. The fact that Iraq identifies itself as a 'radical' state is likely to hold even less sway over the formulation of Soviet foreign policy strategy than it has in previous years. Iraq too is likely to continue to scrutinise the benefits of good relations with Moscow. There is evidence to suggest that the Iraqi regime is in the process of once again courting the ICP. This would improve the atmosphere of bilateral relations slightly, but, more importantly, result in an increase in direct ties between the CPSU and the Ba'th Party. However, the apparent new mood of reconciliation in Baghdad probably owes more to domestic political factors, not least the desire once again to isolate the troublesome Kurdish opposition groups, with which the ICP has maintained close ties, than it does to foreign policy motivations.

For the foreseeable future it seems that the general continuity in Soviet-Iraqi relations will be maintained. This is partly because many of the objective factors which underpinned, or at least did not undermine, the relationship still remain in place. It continues to be the case that few disruptive core issues exist. For instance, there are few direct links between the USSR and the Kurdish opposition groups operating in Iraq, and Moscow has shown itself broadly indifferent to the suffering of some of the Kurdish population at the hands of the Baghdad regime. The Soviet Union has not adopted the visible position of apparent concern exhibited by the US and some European states over Iraq's use of chemical weapons against some of the Kurdish areas in the north. It is also true that neither state is faced with an abundance of alternative allies in the region. The incremental improvement in relations between the US and Iraq has been halted by Washington's concern over the use of chemical weapons. Moreover, Israel has been unhappy at the general improvement in relations which has appeared to be taking place. The Soviet Union had its fingers burnt towards the end of 1987 by attempts to cultivate Iran with a view to acting as a peace broker. Since then Moscow's expectations have been more realistic.

The one factor which some commentators feel may undermine the basis of Soviet-Iraq relations is the direction of Iraqi economic reform. As we noted in Chapter 2, Iraq has been engaged since February 1987 upon an economic strategy of liberalisation and commercialisation. In some cases this has involved limited privatisation. The aim behind this reform has been to improve the performance of the economy and to encourage the private sector to play a more vigorous role within it. Although the state continues to assume a high profile in economic activity, it is now expected to operate in a more commercial way. The demands and expectations of the reconstruction process are likely to add to this imperative. The philosophical basis of the new direction in Iraqi economic affairs would tend to suggest that Baghdad will increasingly be tempted to look towards the West. The private sector is more likely to favour the US, Europe and Japan for the importation of brand-name goods. The state is likely to be tempted by the attractiveness of more advanced technology. If generous financing arrangements were also made available, the impetus of economic utility might increasingly draw Iraq away from its relationship with the Soviet Union. For all that, it is difficult to imagine a fundamental reorientation of Iraq towards the West and away from the eastern bloc. The continuation of Moscow as Iraq's premier source of arms seems likely to continue into the next decade, regardless of changes in the country's broader trading profile.

The US and the Gulf

The Iran-Iraq war never conformed to the bi-polar model which has tended to characterise regional disputes over the last 30 years. The absence of diplomatic relations between the United States and either combatant for the first four years of the conflict is ample demonstration of this. Since 1979, the relationship between the US and Iran has been very poor and often non-existent. The loss of a major ally as a result of the Iranian revolution caused deep resentment in Washington. The brooding suspicion on both sides was transformed into open hatred with the seizure of the American embassy hostages only months later. Even well-meaning but pitiably naive attempts to develop links with 'moderates' in Tehran, such as the Irangate affair, were ill-conceived and ended in humiliation and failure. Until 1984, diplomatic relations between the US and Iraq remained severed. Their resumption, however, owed more to the United States' negative perception of Iran than to a genuine desire to re-establish normal ties with Baghdad. The provision of satellite assistance to the Iraqi military was also more a function of the narrow objective of halting the Iranian war effort than a product of greater co-ordination in military and wider affairs.

In the aftermath of the ceasefire, Washington's attitude towards both regional powers remains tentative. The days of deep bitterness against Iran have gone. The relationship has stabilised in so far as there is a continuing dialogue of an indirect nature. The two sides are engaged in the slow process of confidence building and a slow but steady improvement in bilateral relations looks to be on the cards. There appears to be a sincere desire on the part of the State Department and the pragmatic foreign policy decision-makers in Tehran to see the relationship improve. Direct talks and the establishment of full diplomatic relations look set to take place, although the timescale over which they occur will depend on a number of factors, not least the public perceptions in both countries. The release of the US hostages in Lebanon could also emerge as a *sine qua non* of the restoration of diplomatic relations. Despite a desire on the part of the Iranian authorities to see links improve, the US will do well to conduct its relations with Tehran in a deft and sensitive way. Post-revolutionary Iran is likely to remain suspicious of US interference for many years to come.

It is arguable, however, whether there is much likelihood of an improvement in US-Iraqi relations. Many observers believe that they have progressed as far as they can. Obstacles exist on both sides. Iraq is extremely distrustful of the US. This is based on the experience of bilateral relations over the past 30 years, the ready involvement of the US in the internal affairs of other Arab and Third World states, and the extent to which Israel and its friends have influence over US foreign policy-making. Most recently, Iraq has been suspicious of the role which Israel and its allies on Capitol Hill have played in stoking the engine of anti-Iraqi feeling over the use of chemical weapons. Iraq appears genuinely to think that the outcry in the US in the aftermath of the ceasefire over its use of such weapons in parts of northern Iraq has become more vocal than after other incidents in the past. This Baghdad ascribes to political reasons, interpreting it as being part of the general dissipation of US support for Iraq as the Iranian threat recedes in the aftermath of the war.

There are many on the US side who believe that Washington's relationship with Baghdad has developed to its full, though stunted, potential. These voices are even to be found in the State Department among personnel working on Iraq. They believe that the presence of an organised and influential pro-Israel lobby in Washington makes the chances of advanced weapons sales to Baghdad unlikely in the extreme. However, the pessimism at the prospects for the relationship run deeper. The sceptics point to the nature of the Iraqi regime as having little in common with that in the US, in relation to either ideology or style of government. In particular, anxieties are expressed over the human rights record of the Iraqi regime and its general attitude towards civil liberties.

The Superpowers Interaction and the Region

In theory at least, tremendous potential exists in the Gulf region for an understanding between the superpowers, especially over their respective relations with Iran. Since the overthrow of the Shah, neither Moscow nor Washington has enjoyed an 'insider' relationship with the revolutionary regime.

Indeed, both have had their fingers burnt in trying too quickly or too fundamentally to promote political ties with Tehran. The Soviet Union have been said to enjoy a slight advantage in the closing weeks of the Reagan presidency because it has diplomatic relations with Tehran, and the US does not. Moscow can at least talk directly to Tehran. However, as has been argued earlier, the existence of relations is ironically evidence of the more profound suspicions and deeper complexities which exist in Soviet-Iranian relations.

In relation to Iran, both superpowers share a mirror-image goal as their most important objective: namely that the other superpower should not develop the sort of close and exclusive relationship that prevailed during the Shah's time between the US and Iran. Given this commonality of interest, the potential therefore exists for the two to come to an agreement, tacit or otherwise, that neither will try to pursue such an exclusive insider relationship. For the Soviet Union, this would involve simply putting into practice that element of its new thinking which recognises that the US has foreign policy interests in the region. Indeed, for Moscow, after its experience of the discomfort of the Iranian-US axis of the 1960s and 1970s, the establishment of such an understanding represents the chance to consolidate the break in strategic ties between Iran and the US which occurred with the success of the revolution.

For the US, such an arrangement may prove harder to swallow, primarily because it would represent the final institutionalisation of the sharp reversal in fortunes which it experienced with the downfall of the Shah. While this may be unpalatable, it would only amount to a formal recognition of the state of affairs that actually exists. The fact is that however receptive the Iranian regime may become to assistance from the West in its reconstruction efforts, it plainly does not want the claustrophobic bilateral relationship of pre-revolutionary times. US acceptance of friendly relations between Moscow and Tehran should in theory be easier, given developments elsewhere in the region. The withdrawal of Soviet troops from Afghanistan, assuming it takes place, ought to allay Washington's worries about Soviet expansionism. If, as some Soviet officials maintain, there is indeed no longer a debate in the US over the USSR's participation in an international conference on the Arab-Israeli dispute, then it would appear that the US has already conceded the principle of superpower co-operation in order to act as broker for peace in the wider region. In fact it should actually be easier for the US to accept a diplomatic relationship between the Soviet Union and Iran than a role for Moscow in the Arab-Israeli dispute, if only because in the latter case the Likud bloc in Israel is set against any form of international peace conference, with or without Moscow.

It may also be hoped that, in the climate of detente, the US may come to appreciate that the Soviet Union has legitimate interests close to its borders. The US would do well to consider the parallels between the Soviet Union and Iran, and its own relationship with Mexico. In both cases the superpowers are located alongside important regional powers which are highly conscious of this proximity. In both cases the demarcation between the two sets of states is blurred by the existence within the superpower of large numbers of people who have a strong affinity with the regional power. In the same way that the US regards it as a high priority to cultivate good relations with Mexico City, so one might expect it to understand Moscow's desire to develop normal relations with Tehran.

Despite all the potential for an understanding between the superpowers over Iran, such an understanding is unlikely to take place. On the contrary, the scene could well be set for an increase in superpower competition in the littoral Gulf area. The principal reason why such a favourable scenario is unlikely to occur is the continued existence of the US strategic interest of blocking Soviet expansion southwards. This aim persists unamended, despite the 'new thinking' in the Kremlin. The idea of an understanding with the Soviet Union which might notionally increase the quality of Soviet relations with Tehran would meet stiff opposition in Washington, especially in defence circles. Moreover, there is a rationale for the United States to pursue improved relations with Iran regardless of the Soviet Union, if the analysis propounded in Chapter 1 is accepted as valid, namely, that greater potential exists for a qualitative improvement in US-Iranian relations than it does for US-Soviet relations. In other words, the US might conclude that an agreement with Moscow for broad parity in bilateral relations with Iran would be to sell short the potential for US-Iranian links. This argument is reinforced by the perception which the US has of Iran as a large and growing market backed up by a high degree of effective demand for imported goods. The US will want to establish itself as one of the foremost suppliers of that market. Though the US would expect to have to share the Iranian market with European and Japanese exporters, it could be expected to feel unhappy about the Soviet bloc participating in the supply of goods and services.

If the US rejects the idea of an understanding with the Soviet Union and seeks to maximise its ties with Iran at Moscow's expense, it could discover that this is a strategy containing many dangers. The first is that Iran will play off the superpowers against each other with growing ease. This in turn will contribute to heightening superpower competition in the area. Secondly, should the US meet with success in such a strategy, the danger will be that Washington will overplay its hand. The temptation will be once again to attempt to forge the sort of strategic relationship which existed during the time of the Shah. Such a policy would be misguided and most probably counter-productive. It would reawaken feelings of *déjà vu* from the 1960s and 1970s. It might also help to undermine the position of the pragmatic decision-makers in Tehran. Ultimately, it would almost certainly oblige the Iranian authorities, whoever they might be, to seek improved ties with the Soviet Union.

The potential for superpower competition is not confined to Iran. American fears over Soviet penetration could be re-kindled if Moscow were to improve the quality of its relations with the GCC states. The most important of these is Saudi Arabia. The Soviet Union has not had an ambassador in Riyadh since 1938; indeed, many people forget that the two states ever enjoyed formal links. Furthermore, the somewhat oversimplified impression prevails in the US, even in some goverment circles, that Saudi Arabia is an intensely Islamic country for whom the idea of diplomatic relations with an atheist state is anathema. It is true that the Soviet Union is unlikely ever to be the natural ally of the Saudis, or the eastern bloc their premier trading partner. However, the dangers to Saudi Arabia from regional powers like Israel and Iran have eclipsed the perceived threat from the Soviet Union. Riyadh now appreciates the diplomatic importance of the Soviet Union as a superpower, as the high profile visit to Moscow in January 1988 of the Foreign Minister, Saud al-Faisal, illustrates. Indeed, it seems that it was Saudi Arabia that requested the visit, not the Soviet Union[16]. The exceedingly close relationship between the US and Israel, and the feeling even in the moderate Arab camp that the US cannot play the role of an impartial broker in an Arab-Israeli peace, further emphasises the importance of a central diplomatic role for the Soviet Union in the region.

The succession in Saudi Arabia could also see Riyadh seeking to develop a more even-handed foreign policy. King Fahd has been extremely well disposed towards the US, in spite of periodic rebuffs over arms sales. Indications are that his successor will try to forge a more independent foreign policy. Should Prince Abdullah, the heir apparent, succeed, this would certainly appear to be likely. This is not because Abdullah is at all anti-Western. Indeed, his successful visit to Washington in October 1987 did much to reassure the US on this point. However, there is believed to be general dissatisfaction in the Kingdom at the closeness of the relationship with Washington and the distance between Riyadh and Moscow. Abdullah may be expected to respond to these views. Moreover, Prince Abdullah, or whoever succeeds, will probably not want simply to carry on the policies of his predecessor. There will be at least a superficial desire on the part of the new king to stamp his personality on the foreign policy of the state. If Abdullah does succeed, there are more personal reasons why he might elect to follow a middle course. He is renowned for the good relations which he has cultivated with the Syrian regime. If he is to continue as the conduit through which the Saudi regime maintains contacts with Syria, then a more balanced relationship with the two superpowers appears likely.

State Department officials in Washington recognise that the chances of diplomatic relations being established between Moscow and Riyadh in the wake of the Soviet Union's withdrawal from Afghanistan are high. It is admitted that because of the stereotypical view of Saudi Arabia such a move would be greeted with surprise, even in Washington. Anxiety about growing Soviet influence would be fuelled by the establishment of an embassy in Riyadh and the inevitable access which Soviet diplomats would have both to the palace and the foreign ministry. This in turn might arouse suspicions that the Soviet Union was making inroads through diplomatic means where military means have failed. It would increase suspicions as to Soviet intentions on the part of the US authorities, and this would fuel the lobby for US policy in the Gulf to be aimed more precisely at blocking the extension of Soviet influence in the area.

Conclusion

The 'new thinking' on foreign policy emanating from the Kremlin has major implications for Soviet diplomacy in the Gulf. In particular, Moscow's actions are increasingly not the policies of a state which has a bi-polar view of the world. The US, however, is unlikely to prove quite so amenable to change. Washington's policies in the Gulf are based upon strategic interests which, by and large, have endured over forty years. The US will find it difficult to abandon its view of the Gulf as an arena for superpower competition. It is, therefore, likely to retain as its chief objective in the area the blocking of Soviet expansion southwards.

Regardless of what might be happening on the central issue of superpower detente, the potential for the Gulf to become the focus of regional competition between the US and USSR could increase. This would be facilitated by the re-establishment of diplomatic relations between Moscow and Riyadh, in the aftermath of the former's withdrawal from Afghanistan, and the likelihood of growing attempts by both Moscow and Washington to cultivate Tehran. For all that, the potential for dramatic re-alignments in the area is limited. The close proximity of the Soviet Union to Iran, combined with the unfavourable experience of Iran with both the Russian Empire and the Soviet Union, severely reduce the potential for an improvement in links. On the other hand, the nature of the political system and the negative view of Israel make it unlikely that US-Iraqi relations will move much beyond the current plane. As a result, the present level of Soviet-Iraqi relations, based as it is on no more than mutually perceived utility, looks set to continue to prosper, while US-Iranian relations, when unencumbered by such issues as the hostages in Lebanon, could, in the course of the 1990s, come increasingly to mirror Soviet-Iraqi ties.

Notes

[1] Interview with senior official in Tokyo.

[2] For example see Philip Robins, 'A Feeling of Disappointment: the British Press and the Gulf Conflict', *International Affairs*, Vol.64, No.4, Autumn 1988.

[3] 'US Policy in the Persian Gulf', *Department of State Bulletin*, October 1987, p.38.

[4] Ibid.

[5] President Carter's State of the Union address delivered one month after the Soviet invasion of Afghanistan.

[6] 'US Policy', op. cit.

[7] Sir Curtis Keeble, 'British Policy - The Soviet Union and Eastern Europe', unpublished paper delivered at a conference on: 'The Future of British Foreign Policy', University of Wales, Gregynog, 3-5 November 1988.

[8] See *The Economis*t, 18-24 June 1988.

[9] Steven Sestanovich speaking at a Centre for Strategic and International Studies conference, Washington, DC, 13 June 1988.

[10] For details, see Dilip Hiro, *Iran Under the Ayatollahs*, London, Routledge & Kegan Paul, 1985, p.280.

[11] In February, 1987. Quoted in *Soviet Union and the Middle East*, Vol. XXII, No. 2, 1987, p.14.

[12] Ibid.

[13] Yuli Vorontsov, quoted in *Soviet Union and the Middle East*, Vol. XXII. No. 8, 1987, p.5.

[14] For instance Britain accused the USSR of failing to draft an arms embargo resolution as earlier promised; see *The Guardian*, 13 November 1987.

[15] For instance on 29 November 1987 the Iraqi Foreign Minister, Tariq Aziz, stated in an on-the-record interview that the position of the Soviet Union was 'an obstacle' to the adoption of a follow-up resolution. He went on to say that other Arab states were pressurising the Soviet Union on this point and even criticising it outright.

[16] Soviet Foreign Ministry spokesman Gennady Gerasimov stated that the visit took place at the request of Saudi Arabia. Quoted in *The Financial Times,* 27 January 1988.

5. THE IMPORTANCE OF OIL SUPPLIES

by Jonathan Stern

Introduction

The geographical concentration of low-cost oil reserves in the countries of the Gulf is a major, perhaps the major, reason for the strategic importance of the region. Over the past two decades, and particularly in the period since 1973, there have been periodic reminders of this fact as domestic and foreign policy actions by regional governments, as well as incidents within and between countries, have combined to affect levels of oil production and exports, and hence world oil prices, with sharp repercussions on the world economy.

Many studies have been produced covering these subjects over the past two decades. The aim of this chapter is therefore to examine the changes in attitudes towards the importance of Gulf oil in the wake of two events: the fall in world oil prices (which occurred in 1986) to less than $20 per barrel, which seems likely to continue for a period of years; and the apparent end of the Iran-Iraq war in 1988. To this end, the chapter is divided into three sections: a consideration of the situation in oil-exporting countries in the 1980s; the effect of the Gulf war on oil exports; and the 'new conventional wisdom' on world oil supplies and prices. Finally some conclusions are offered regarding the likely importance of the Gulf in world oil trade by the end of the century.

A large number of factors determine the capacity and willingness of a country to produce and export oil. Although resource endowment is the most obvious attribute, it is not the only factor and may not even be the most important determinant of production and export policy at any particular point in time. Having established an adequate resource base, governments may be equally preoccupied with: revenue requirements, debt-servicing requirements, overseas development plans, protection of future markets (and/or supporting current price levels), domestic social and political factors, and foreign policy considerations[1]. Because the major focus of this chapter is on the importance of the Gulf region in world oil trade, the following sections are therefore concerned primarily with capacity rather than willingness to export; and with export rather than production potential.

Oil-Exporting Countries in the 1980s[2]

The importance of the Gulf

Table 5.1 shows the oil reserves of major oil-producing countries at the end of 1987. The principal Gulf producers, Saudi Arabia, Iran, Iraq, Kuwait and the UAE, account for more than half of total world proven reserves and more than three-quarters of the reserves of OPEC member countries[3]. However, proven reserve figures for Gulf countries, particularly for Saudi Arabia, Kuwait and the UAE, are likely greatly to understate the potential resource base. Proven reserves reported by the Gulf governments increased significantly during 1987 and there is little doubt that ultimately recoverable reserves in these countries far exceed the proven figures[4].

From Table 5.2, it is evident that in 1987 Gulf oil producers no longer dominated world oil production, but as the figures for production capacity in Table 5.3 demonstrate, this was not because the Gulf had lost the ability to produce larger quantities of oil[5]. While the figures for production capacity are subject to some uncertainty, the general trend is clear: the Gulf countries are producing at well below even the capacity which could be made available at short notice. The clearest example is Saudi Arabia where production of 4 million barrels per day (mmbd) in 1987 compares with a sustainable production capacity of approximately 6-7 mmbd and a sustainable capacity which could be mobilised with modest investments of 8-10 mmbd[6].

Other countries in the region are in a similar position, albeit on a smaller scale. Production constraints have been imposed on all members of OPEC (with the exception of Iraq) since 1982, in order to sustain the price of oil, although some, such as Kuwait, have limited production since 1973. Over the past decade, Iran and Iraq have experienced special production and export problems. Iranian production facilities were damaged as a result of the war, and during this period it is unlikely that the equipment and expertise to implement the complex gas reinjection programme necessary to prevent a decline in recoverable reserves were available[7]. In the 1980s, Iraqi production has been limited due to inability to export oil via the Gulf. Nevertheless, the construction of pipeline networks (see Map 2) during the 1980s has allowed the expansion of Iraqi exports. Production from the six Gulf members of OPEC in 1987 amounted to 11.7 mmbd, compared with immediate available capacity of 17-20 mmbd and an ultimate potential of 20-25 mmbd (Table 5.4).

The Gulf members of OPEC

Oil analysts spent much of the 1970s writing about the dominant position of OPEC producers, particularly those in the Gulf. World oil trade statistics (Table 5.5) are a reminder of how the world held its breath at each OPEC meeting particularly during the period 1973-80. This was the time of statements that oil was 'running out' and doubts as to whether OPEC would be willing and able to produce 40-50 mmbd during the 1990s in order to meet expected demand[8].

In the event, OPEC oil production and exports peaked in 1977 at 31.2 mmbd and 29.4 mmbd respectively, but fell rapidly in the 1980s, to a low of 15.4 mmbd and 13.2 mmbd in 1985, as the exporting countries attempted to hold prices at levels exceeding $30 per barrel which had been established after the second major world oil price rise in 1979/80[9]. Inability to hold prices at such high levels, or (many would argue) the result of these high prices, was graphically demonstrated by the price collapse of 1986. By the following year, OPEC export levels had recovered somewhat to around 16 mmbd, but abundant availability of oil supplies from a range of countries, OPEC and non-OPEC, prevented any sustained price recovery from the short-term equilibrium level of $15-18 per barrel, and constantly threatened to force prices down to much lower levels. In the period 1986-8, these events caused world oil and energy watchers radically to reassess their views of both the current situation and future trends.

The reasons for the decline in demand for OPEC oil and the fall of oil prices are complex. On the supply side, the price rises of 1973/4 and 1979/80 brought forward a large quantity of additional oil production, from both OPEC and non-OPEC countries. During the 1970s, the Gulf states alone accounted for 60 per cent of the crude oil in world trade, whereas by 1987 that figure had fallen to 40 per cent (Table 5.5). The move away from dependence on OPEC oil was reinforced by diversification into non-oil fuels, principally natural gas, coal and nuclear power.

On the demand side, recession and industrial restructuring away from energy-intensive industries, combined with greater efficiency of energy use and energy conservation, led to a rapid and sustained reduction in the energy/GDP ratios of all OECD countries. Although world energy demand in 1987 was 32 per cent higher than its 1973 level, oil demand was only 3 per cent greater and the share of oil in energy balances had fallen from 47 per cent to 38 per cent (Table 5.6). Looking at OECD countries over the same period, both effects are magnified with overall energy demand only 6.3 per cent higher, and the share of oil reduced from 53 per cent to 43 per cent [10].

With reduced demand for oil and a growing number of oil-exporting countries, it was not surprising that, as the 1980s progressed, OPEC struggled increasingly to maintain its grip on oil prices. With the need to impose progressively more severe production quotas in order to support prices, discussions within the organisation became increasingly acrimonious. By the mid to late 1980s, two broad price policy groupings could be identified within the organisation: one led by Saudi Arabia and the countries of the Gulf Co-operation Council (GCC), in favour of prices sufficiently low to stimulate demand and hold back non-OPEC production; the other led by Iran pressing for higher prices (generally defined as in excess of $20 per barrel) in order to boost revenues[11]. Following the 1986 oil price collapse, OPEC meetings tended towards some compromise between the two groups, usually proving unsatisfactory to one or other, reflecting as much political strength as commercial strategy. When looking at the more general influence of Gulf countries on oil prices, it is clear that the main Gulf OPEC countries, Saudi Arabia, Kuwait, the UAE, Iran and Iraq, continued to be the key OPEC players, despite the fact that they were divided in their views. Thus as far as short-term oil price policy was concerned, the 1980s saw a sharp division within the Gulf countries between Iran and the Arab Gulf countries, reflecting the political divide of the Iran-Iraq war.

Comparing the Gulf countries with all other oil exporters, in the longer term, their dominant position is obvious. With huge reserves being produced at a relatively slow pace in comparison with their potential, it is not surprising to find that reserves to production ratios in Saudi Arabia, Kuwait, Iran, Iraq and the UAE comfortably exceed those of all other countries (Table 5.1). Whereas these countries can continue to produce and export for more than a century at the levels of the late 1980s - or even at the much higher levels of sustainable production capacity indicated in Table 5.4 - the reserves of most other producers are in the range of 10-40 years. In addition, whereas the domestic consumption of the Gulf countries is relatively low so that the distinction between production and exports becomes blurred, this is not the case in a large number of the other oil-exporting countries where rising domestic consumption may threaten export surpluses over the next decade.

Non-Gulf members of OPEC

In terms of reserves and future export potential, by far the most important OPEC member country outside the Gulf is Venezuela. Venezuelan proven reserves have more than doubled since 1985, reaching 56 billion barrels in 1987[12]. New discoveries are likely to add another 9 billion barrels to reserves between 1988 and 1992[13]. With this resource base Venezuela would have the ability to double export volumes from the 1987 level of 1 mmbd. Elsewhere among OPEC members, the strongest position is held by Libya, with an appreciable resource base and a comparatively small population. Again there seems little reason, on resource grounds, why Libya should not significantly increase its oil exports over the next decade. Algeria, Nigeria and Indonesia, with smaller resource endowments and very large populations (particularly in the latter two countries) are in a somewhat weaker position. Any prediction for Nigeria has to balance increasing domestic demand against the likelihood of discovering substantial additional reserves. Algeria and Indonesia, with rising demand and smaller reserves, may experience some decline in exports over the next decade.

Other exporting countries

Of the many oil-exporting countries outside OPEC, undoubtedly the most important in the longer term will be Mexico. Mexican resources indicate the potential for significant increases in export volumes from the 1987 level of 1.35 mmbd[14]. The country's large population and severe economic and financial problems provide compelling reasons to expand exports in the future. Mexican solidarity with OPEC price support policy, combined with a reduction in funds available to Pemex, the national oil company, has prevented this course of action being followed over the past several years.

In the North Sea, the rise of the United Kingdom and Norway as oil exporters of considerable significance - with net exports of 0.98 and 0.80 mmbd respectively in 1987 - has been one of the major features of the 1980s. While predictions are always fraught with difficulty, there is some agreement that British oil production will decline relatively rapidly in the late 1980s and early 1990s, from 2.35 mmbd in 1987 to 1.65 mmbd in 1990 and possibly to 1.1 mmbd by 1995[15]. This would mean that, before 1995, the UK would cease to be a net oil exporter. The Norwegian situation is considerably more optimistic, with the Ministry of Petroleum and Energy suggesting that '..oil production will increase by about 60% ...up to the mid 1990s..(and)..might increase to twice the level of (1987) production'[16]. Norway might therefore increase exports from 0.8 mmbd in 1987 to well over 1 mmbd by the mid-1990s.

Soviet oil exports to world markets, which a decade ago were almost universally expected to decline during the 1980s, have increased strongly, reaching 1.7 mmbd in the late 1980s[17]. Of all the major exporters, it is most difficult to predict future Soviet export levels. Although there is every indication that Soviet exports will continue at current levels in the short term, there are major differences of opinion about whether the 1990s will see a slow increase or a slow decline in this figure[18]. In China, disappointing offshore exploration combined with ambitious industrialisation plans (suggesting the need for much greater domestic consumption) make it uncertain whether exports can be maintained at the 1987 level of 0.5 mmbd[19].

Table 5.7 groups a large number of smaller oil producers under the headings: importers, exporters and self-sufficient countries. Collectively these 20 countries produced more than 7 mmbd in 1987. Classifying a large number of countries from different geographic regions, at different stages of development, in terms of their oil trade position is useful only in the cause of brevity. While some of those currently self-sufficient - such as Tunisia and Peru - are in danger of becoming importers, a number of others: Oman, Syria, Angola, and Colombia, may enhance their position in the future. Perhaps the most important country not listed in Table 5.7 is North Yemen which may be producing as much as 0.5 mmbd in the 1990s[20].

The cost of oil production

Decisions regarding the speed and extent of oil exploration and development are a function of the cost of these activities compared with the current and expected international price. It is important to make the distinction between exploration and development costs for new fields and operating costs for existing fields. When world oil prices fell below $20 per barrel, a number of exploration and production programmes came under threat; few operating oil fields anywhere in the world would be threatened unless prices fall to $5 per barrel.

In addition to their enormous resource base, the Gulf producers have the advantage that the cost of oil exploration and development in their countries is extremely low (Table 5.8). Although generalisations are difficult, because of differences between fields and exchange-rate fluctuations, in the late 1980s few new Gulf oil fields cost more than $1 per barrel to develop, with the average cost being in the range of 30-60 cents per barrel. This compared with American production costs of $5-10 per barrel for new fields in the Lower 48 States and around $15 per barrel for Alaska. In the UK North Sea, production costs for new oil fields in 1987/8 were in the range of $15-20 per barrel, with Norwegian sector costs being slightly higher.[21].

When the oil price fell from a range of $25-30 to less than $20 per barrel in 1986, it was expected that oil exploration and development in a large number of non-OPEC countries would fall sharply and would not recover until prices rose again. The trend in the United States was both untypical of other non-OPEC countries and less severe than had been expected. Oil production fell by 0.31 mmbd in 1986 (compared with 1985) and by 0.33 mmbd in the following year, but there is an expectation that the decline over the next several years may be smaller, perhaps in the order of 0.1-0.2 mmbd[22]. A more pessimistic assessment, made in the aftermath of the 1986 price collapse, was that, with world oil prices remaining in the range of $12-18, US production would fall to around 6-7 mmbd by 1990 and 5-6 mmbd by 1995, compared with 9.0 mmbd in 1985[23].

Elsewhere the picture seems less gloomy: in the North Sea, a combination of technical and financial ingenuity saw a (reduced) number of new field developments going ahead at prices in the $15-18/bbl range[24]. The most startling example was the decision, announced at the lowest point of the 1986 price collapse, to proceed with the huge Norwegian Troll gas development. In general, North Sea gas exploration and development, in both the Norwegian and UK sectors, has proved more resilient than oil.

Despite the fact that there has been an undeniable reduction in exploration and development, given two years of sharply reduced oil prices and the expectation of several more years of similar prices, the levels in activity in many countries appeared more robust than expected. The reasons for this require a country-by-country explanation and involve some mixture of: improved technology and engineering, reduced costs of exploration and development (partly as a consequence of advances in technology), reduced expectations of profitability on the part of operators, and a more attractive fiscal regime created by governments.

An intriguing question for the next decade is whether the costs of exploration and development can continue to be reduced if oil prices remain at around $15 (1987) per barrel. A study of non-OPEC reserves (roughly one-third of total world oil reserves) suggests that at an oil price of $10 per barrel only one third of such reserves would be economic to produce (a large share of these in Mexico)[25]. If there is a generalisation which can be made across a large number of countries, it is that at oil prices of less than $20 (and particularly less than $15) per barrel, the OECD oil producers and exporters, the USA, the UK and Norway, have been and are likely to be hit harder than the developing countries. This could begin to have a significant effect if these levels of oil prices continue into the 1990s.

The Effect of the Gulf War on Oil Supplies.

In the period 1984-8, the evolution of the Gulf conflict from a land war to hostilities against Gulf shipping - the tanker war - refocused world attention on the conflict, as superpower and West European navies moved into the region to protect the passage of shipping. However, the remarkable feature of the tanker war phase of the conflict was that it had little sustained effect on flows of oil from the Gulf through the Strait of Hormuz. Comparing the years 1984 and 1987, attacks on oil-related installations rose from 43 (including 39 tankers) to 155 (135 tankers)[26]. Despite the fact that these attacks led to occasional loss of cargoes and loss of life, merchant shipping showed a continuing willingness to run the risks. At the same time, the presence and determination of foreign navies in the region demonstrated conclusively that neither of the belligerents could pose a *sustained* threat to the passage of oil through the waterway[27].

Given a decade of literature describing the Strait of Hormuz as a vitally important 'global choke-point', and that most commentators agreed with US

Secretary of Defense Weinberger that 'The umbilical cord of the industrialised free world runs through the Strait of Hormuz into the Arabian Gulf and the nations which surround it..' , it came as something of a surprise to discover that the passage of oil supplies through the Strait was not seriously disrupted even under conditions of severe regional instability[28]. Even more surprising during this period was the relative strength of world oil prices in the face of what could be described as 'worst-case' military scenarios being played out in the Gulf. Although periods of major naval tension and successful attacks on shipping caused increases in spot prices on a daily or weekly basis, the overall impression was that, in the mid to late 1980s, the conflict did no more than lend some support to prices in a chronically depressed market.

Moreover, the relative strategic importance of the waterway declined as increasing quantities of oil began to reach markets by pipeline. One of the effects of the war was to accelerate the construction of overland pipeline outlets for oil. This process is well advanced as shown in Map 2. In the 1970s, of the 18-20 mmbd of oil exported from the Gulf, less than 10 per cent - roughly 1-2 mmbd of Iraqi oil - was transported through two pipelines: one via Syria and Lebanon, the other via Turkey. By the mid to late 1980s, not only had Gulf exports fallen to less than half their previous levels, 8-10 mmbd, but pipeline capacity out of the Gulf had risen to more than 4.0 mmbd. This is split between two routes: one across Saudi Arabia to the Red Sea port of Yanbu, carrying Saudi and Iraqi crude; and the pipeline system between Turkey and Iraq (see Map 2 and Table 5.9). Iraq has built a new pipeline (to be commissioned in late 1989) which runs parallel to the Saudi Petroline, which has increased its pipeline export capacity to some 3.2 mmbd. Saudi Arabia is planning to expand Petroline (possibly to provide capacity for GCC use).

In the late 1980s, there was a great deal of discussion, but very little action, regarding possible Iranian oil pipeline outlets. Aside from the possibility of a line through Turkey, the conversion of the Iranian gas (IGAT) pipeline to carry oil to the USSR and thence to Western Europe received serious attention[29]. In 1988 Iran started construction of a pipeline with a capacity of 0.6 mmbd from Gurreh to the port of Taheri, although the plan to extend this to Jask, beyond the Strait of Hormuz, seems less certain[30]. Thus Iraq is clearly in a much better position than Iran which, in the late 1980s, has barely got its pipeline plans off the drawing board. In the period up to the early 1990s, therefore, Iran is likely to remain much more vulnerable than Iraq to incidents involving the passage of oil tankers through the Gulf and the Strait of Hormuz.

All indications are that the Gulf countries intend to expand the pipeline networks which would enable them to move a steadily larger proportion of their exports by routes avoiding the Gulf. While these lines fall some way short of a total non-maritime marketing solution for Gulf oil, they provide an immensely important alternative, particularly for Iraq, Saudi Arabia and possibly other GCC countries. The construction of pipelines with reversible flow capability within Iraq, enabling oil from the fields in Kirkuk and around Basra to be exported either via Turkey or via Saudi Arabia, added greatly to Iraqi flexibility in this respect[31].

During a war, pipelines can generally be repaired relatively quickly (in comparison with maritime oil export facilities) following missile attack, and are hence a great asset. However, they are extremely vulnerable to political disagreements between countries, and with the ending of the war the urgency of pipeline construction plans may be slowed as disagreements cause delays in construction[32]. The lines which have been built may shift the focus of attention to Iraqi relations with Saudi Arabia and Turkey. The growing importance of pipeline routes through Saudi Arabia will greatly increase the strategic significance of the Red Sea and Suez Canal as oil waterways[33]. However, it should not be forgotten that, in the past, changing political allegiances have meant that some regional pipelines have lain idle for protracted periods of time[34].

The end of the war

During the latter part of the 1980s, it became a commonplace to remark that in military/political/strategic terms all countries, both regional and external, supported an end to the war, whereas, in terms of oil price trends, all OPEC and many OECD countries greatly feared such a development. Despite the difficulties for OPEC in regulating production quotas and the periodic concerns in OECD countries about security of supply, the curtailment of oil exports from the belligerents lent support for prices in a chronically depressed market. In the absence of a major regional war, the oil price collapse of 1986 might have occurred some years previously[35].

In 1989, assuming that the war is substantively over, two scenarios have been advanced regarding oil supplies and prices. Under conditions of prolonged ceasefire, with or without a peace treaty, both belligerents will embark on major rebuilding programmes requiring massive additional revenues. Such developments give rise to expectations of a surge in oil exports from both countries which would cause great problems within OPEC and place severe downward pressure on world oil prices. Conversely, it has been argued, greater political cohesion within OPEC, which is assumed to result from the end of the war, will make agreement on production quotas easier to reach and will result in a greater effort to support higher price levels, particularly given the stated Iranian desire to raise oil prices to a level around $20/bbl.

While there can be no great degree of confidence about a consistent trend over a prolonged period, it seems likely that Iran and Iraq will need to rebuild both their civilian and military infrastructure for a period of at least five years. There will be strong pressures to increase oil exports to support this reconstruction. In such a situation, continued and even intensified strains within the OPEC structure of production quotas can be expected to continue. The result is likely to be continuing pressure from both Iran and Iraq to expand oil exports, with or without the agreement of other OPEC members, thus exerting downward pressure on world oil prices in the short to medium term.

In the unlikely event that serious hostilities between the countries should resume, this would once again provide support for oil prices. In terms of export volumes, future military activity would be most likely to affect Iranian sales adversely since, as mentioned above, large-scale pipeline construction would allow Iraqi oil exports to expand to 3-4 mmbd by the end of 1989.

The 'New Conventional Wisdom'

The reduction of the share of oil in energy balances and the fall in the world price of oil in 1986 have caused a fundamental rethink about the future role of oil and the importance of OPEC (specifically Gulf) supplies within the global oil scene. Prior to the 1986 oil price collapse, the 'conventional wisdom' suggested that the OPEC share of world oil supplies had reached its lowest level, and that, in the early 1990s, a combination of declining non-OPEC supplies and rising demand would once again put upward pressure on prices and increase the power of OPEC in the international market. The oil price crash of 1986 seemed to reinforce this position. It was assumed that non-OPEC supplies would be badly hit, and that oil and energy demand would receive a major boost as a result of lower prices. Two years later, the general perspective is a great deal more optimistic in terms of avoiding a return to the levels of dependence on Gulf and OPEC oil supplies which were seen in the 1970s. The sections above on non-Gulf oil-exporting countries reflect this optimism. The 'new conventional wisdom' suggests that any tightening of world oil markets, characterised by rising real prices and a much greater call on OPEC oil, will be delayed to the late 1990s, or even the early part of the next century, thus suggesting a decade of relative calm in world markets.

For those with an intuitive suspicion of such forecasts, the difficulty is to decide whether the new conventional wisdom is likely to be wrong for the *same* reasons as previous projections, i.e. that the factors under consideration are so difficult to forecast that the task is impossible; or whether it is likely to be wrong for *different* reasons: i.e. that the major factors (or major surprises) of the future are not amenable to forecasting, and that forecasts (in this case of continuing high prices and major shortages) always carry within them the seeds of their own destruction. As far as the next five years are concerned, it will be particularly important to see whether the rather persuasive logic of 'inbuilt responses' to sharp oil price movements proves to be correct.

Given a period of real prices within the range of $15-20 (1988) per barrel, unforeseen events could cause a sharp rise or a sharp fall in prices. In both cases, the effects may be short-lived because of inbuilt response mechanisms established over the past decade. When prices swing up, the forces which act to push them down include: increased worldwide exploration investment, greater substitution of oil by other fuels, greater energy conservation and efficiency. Conversely, when prices swing down sharply, the forces likely to push them back up include: substitution of coal and gas by fuel oil, reduced OPEC revenues (leading to agreement to reduce exports), lower non-OPEC production, reduction in worldwide exploration. These inbuilt response mechanisms did not exist at the time of the first oil shock (1973/4) and were relatively slow to be activated at the time of the second shock (1978/9), but they are now in place and come into operation more rapidly as a result of sharp price movements in either direction.

Response mechanisms can be divided into two sets of categories: supply and demand, short- and long-term. The supply side, both short- and long-term, is crucially dependent on OPEC action with respect to volumes and prices of oil exports. In the short term, OPEC discipline with respect to production and exports is the critical variable. In the longer term, the major issue will be how quickly spare production capacity, particularly in the Gulf countries, will be fully utilised. However, in the late 1980s, demand-side responses are likely to be more rapid (and perhaps more significant) than supply-side responses.

Relying on inbuilt response mechanisms to correct oil price fluctuations gives rise to problems of two kinds: the time lags before these mechanisms take effect and the economic and social costs which may be incurred in the meantime. In the future, time lags are likely to be shorter than the seven years between the 1979 price rise and the 1986 price collapse; however, response mechanisms tend to work better in developed countries and to impose particular costs and hardships on developing countries. In the last analysis, relying on response mechanisms may be no more than the policies adopted by many OECD governments in the 1980s of promoting 'market forces' in the energy sector wherever possible.

Focusing on the demand side of the energy balance, the single most important factor is the projection of OECD (particularly US) economic growth and oil demand. An extra percentage point on economic growth over a fairly short period would have a significant impact on energy and oil demand. Similarly, a sustained increase of one or two percentage points in oil demand, perhaps reflecting the renewed attractiveness of lower priced oil over other fuels, could, within a few years, significantly alter the world supply/demand picture. Given that a very large percentage of any forecast increase in global oil demand is in the United States, the policies of the Bush Administration will be of the utmost importance. Between 1985 and 1987 US imports of crude oil and petroleum products had increased from 5.1 to 6.6 mmbd and by mid-1988 were running at 7.3 mmbd[36]. There is a consensus that, given oil prices at 1986-8 levels and no concerted government action in terms of oil import fees, minimum support prices, gasoline taxes, etc., US oil imports will reach 8-10 mmbd by 1995, due to a combination of falling domestic supply and rising demand[37].

Major increases in oil demand are also foreseen in the developing countries. However, given that these countries account for only some 17 per cent of total world oil demand (compared with over 60 per cent for OECD countries and 22 per cent for the centrally planned economies), current annual increases of 2-3 per cent per annum in developing country oil consumption would have to be sustained for a period of several years before these have more than a modest impact on the global situation[38].

Against this rather 'business as usual' demand scenario have to be set conservation and energy efficiency gains, and particularly the technological advances which are being made in the more efficient use of energy in all markets. Although some energy conservation gains made over the past decade may be reversible at lower oil prices, there is a permanent effect from the experience of the 1970s. Furthermore, the considerable advances in efficient energy-using plant, vehicles and appliances *are* irreversible and many of these effects are still working their way through the system.

Lower oil and energy prices will have a negative impact both on the behaviour of consumers and on the introduction of more advanced and costly technologies. As far as technologies are concerned, lower prices suggest a slower pace of development. Nevertheless, they will not be 'uninvented' and will remain available for a time when energy prices rise to levels which render them competitive. Moreover, as we have seen with supply-side technologies, their costs may fall to the extent that they can be introduced at lower price levels. Demand-side technologies are likely to receive a boost from environmental concerns arising from the burning of fossil fuels, which will emphasise the importance of energy conservation, efficiency and new energy technologies. Although studies which predict the world's ability to sustain economic growth over the next several decades without major conventional supply-side energy growth may prove to be too optimistic[39], there is little doubt that the potential for conservation and efficiency improvements, as well as for the development of renewable sources of energy, has been seriously underrated in traditional energy analysis.

A final point, which may support the new conventional wisdom, concerns changing ownership structures in the world oil industry. A feature of the mid to late 1980s has been the purchases by major oil exporters of refining and marketing outlets in importing countries. Acquisition of large equity holdings in North America and Western Europe by Kuwaiti, Saudi and Venezuelan companies was prompted, to some extent, by a desire to secure markets for oil products during conditions of oversupply. This may suggest the possibility of a different kind of partnership between exporters and importers in the 1990s[40]. Exporters with a major financial stake in the downstream operations of importing countries will have an incentive to ensure both supply to, and profitability of, these operations.

Conclusion

Over the past decade, Gulf oil supplies have become much less important in world oil trade in comparison with their importance in the 1970s. The key issue is whether, in the 1990s, the enormously strong reserve position of the Gulf countries will once again translate into the market power of the 1970s, with all the political, economic and strategic ramifications of that situation.

Table 5.10 is an attempt to project the export capability of current and likely future oil-exporting countries in the period up to 2000. Export capability does not necessarily say anything about *willingness* to export or the *availability of markets* for exports. Table 5.10 is thus not a projection of likely exports but an illustration of the capacity of countries outside the Gulf to challenge the market power of the region by the end of the century.

This gives us a picture of a relatively small number of big oil-exporting countries (BOECs) - four Gulf producers - with present exports exceeding 1.5 mmbd and with the ability to raise this figure by at least a further 1 mmbd. The USSR comes into the BOEC category but may fall away somewhat in the 1990s. On the other side of the table are a large number of comparatively small oil exporters - almost all of which are developing countries - with export capacities less than 0.5 mmbd. Between these two groups lie nine medium-sized exporting countries with present exports of 0.5-1.5 mmbd. In terms of resource and potential, a good case can be made for including Kuwait, Venezuela and Mexico in the BOEC category. These three countries, and to a lesser extent Libya, have the capacity to re-establish their 1970s status as big exporters. Of the rest: Norway will enhance its position but not into the big league; Nigeria is likely to remain at, or slightly below, present levels; Indonesia and Algeria are likely to become less important; and the UK will slip out of the picture altogether.

Classifying exporting countries by the size of their exports has the merit of identifying the most important potential parties to any agreement on volumes and prices, irrespective of their geographical location and political/organisational grouping. A further useful distinction may be made between those countries with large reserves and small populations, which may have some flexibility in their revenue requirements, and those with large populations and low per capita incomes, which have less room for manoevre. In any event, the usefulness of grouping oil-exporting countries under headings such as OPEC, Gulf-OPEC, non-Gulf OPEC and non-OPEC has become less obvious. In the late 1980s, it has become much more useful to categorise countries by the size of their likely exports over the next decade.

The notion of OPEC negotiating with 'non-OPEC' countries, which surfaced in the spring of 1988, is particularly problematic. The fact that these negotiations came to nothing was far less important than the fact that this particular 'non-OPEC' group did not contain the USSR, the UK and Norway, clearly the three most important non-OPEC exporters after Mexico[41]. Although OPEC's performance in coping with the price collapse during the 1986-7 period was more successful than might have been expected, in general over the past decade the organisation has become less effective as a price-setting cartel. To some extent this stems from political disagreements between the Gulf members as market conditions became more difficult. Equally, however, the organisation contains some small exporters which do not assist its exercise of market power, and lacks a number of significant exporters without whose co-operation effective action on prices cannot be taken. If this was merely a question of dealing with three or four additional major exporters the position would be difficult but potentially manageable. However, OPEC faces immense difficulties in coping with a proliferation of small producers and exporters which, while individually insignificant, are collectively important, especially in a weak oil market. The addition of such a large number of small suppliers fatally weakens the already small chances of political cohesion in support of policies requiring stringent production discipline to achieve higher prices.

These trends are likely to give OPEC, as currently constituted, continuing difficulties in the 1990s. However, in the late 1980s, the position was entirely different. In a world of chronic oversupply OPEC - whatever its failings - is the only organisation which can prevent a production and export free-for-all which could give rise to single-figure oil prices for a period of years.

Within the organisation the five big Gulf exporters and Venezuela will remain the key players. Table 5.10 suggests that Mexico and to a lesser extent Norway and the USSR, will remain the most significant, but by no means the only, exporters with which OPEC will have to deal. A key issue in the 1990s is whether the number of smaller exporters, and the quantities of oil which they contribute to the world market, will continue to grow; whether in fact Table 5.10 is correct to place a large number of these countries in the category of 'modest increase' in production in the 1990s. The smaller the number of big exporting countries, OPEC or non-OPEC, and the larger their share of the oil in world trade, the more likely they are to be able to dominate world markets and prices. .

As far as the 'new conventional wisdom' is concerned, there is a danger that such wisdom is, almost by definition, condemned to the same fate as the previous wisdom of the 1970s. Assuming that there is some merit to the new conventional wisdom, at least two interesting questions arise concerning the Gulf countries and their oil policies. Given that the Gulf countries are unlikely to be able to raise oil prices above $20 per barrel for any prolonged period without activating response mechanisms which give rise to major downward pressure on prices, they may decide to keep oil prices low for a protracted period in order to recapture market share. One interpretation of post-1986 Saudi (and Kuwaiti) oil policy is that the

Kingdom is determined to maintain the world price of oil, for a prolonged period of time, at a level which is sufficiently low to stimulate world oil demand and hinder investments in non-OPEC oil and other energy sources. In 1988, that level might be as low as $14 per barrel.

This in turn raises the issue of whether exporters - individually or as a group - have the power to fine-tune prices over long periods within a narrow range. There are also significant short- to medium-term costs involved in keeping prices at these low levels. In 1988, budgetary difficulties saw the Saudi government seeking to borrow money for the first time in many years. Nevertheless, larger exports at lower prices may be a strategy which suits a number of major Gulf oil-producing countries for different reasons: Saudi Arabia and Kuwait because of a desire to promote increased market share in the longer term; Iran and Iraq because of a desire to export greater quantities in order to boost revenues for economic and military reconstruction. While this policy may be less than optimal in terms of revenue maximisation, it may be a politically acceptable compromise during a decade in which excess supply is likely to be an ever-present feature of world oil markets.

The second question which arises from the new conventional wisdom is how the region is likely to be perceived by oil-importing, specifically OECD, countries. Given the experience of two major price increases in the 1970s, it seems worryingly complacent to assume no likelihood of a major oil price and supply dislocation over the next 5-10 years. Major and sustained increases in OECD economic growth, leading to comparable increases in energy, and specifically oil, consumption and imports will raise concern, not necessarily in terms of a major supply disruption (although this remains a possibility), but in terms of growing dependence on oil imports. The problem for many OECD governments, and especially the United States, will be to reconcile their convictions regarding the importance of promoting market forces with the problems of energy security which may be created, specifically in terms of growing dependence on imported oil from the Gulf region. The protection of domestic OECD oil (and to a lesser extent gas and coal) industries will remain an issue as long as world oil prices remain significantly below $18 (1988) per barrel.

The new conventional wisdom does not in any sense suggest that the Gulf is, or will ever be, unimportant in the world oil supply picture. Dramatic regional events involving Arab-Israeli tensions, unrest within the Palestine population, and military conflagration in the Gulf will always give rise to concern about the potential impact on oil supplies and the ways in which access to Gulf oil can be ensured. However, there is considerable confidence both that global energy balances will not be profoundly disturbed for much of the next decade as long as the region contributes 10-12 mmbd to world oil trade and that there are very few circumstances in which this is unlikely to be the case since:

a) the regimes in these countries, irrespective of ideology or religious conviction, will need a minimum level of revenues from oil exports.

b) whatever military/political conflict occurs within or between countries, there are an increasing number of routes by which oil can leave the region, notably overland routes which reduce the strategic importance of the Gulf as a waterway.

Specific countries within the OECD region, particularly in continental Europe and Japan, have been, and are likely to remain, heavily dependent on Gulf oil. Nevertheless, the major conclusion to be drawn from this analysis is that at currently expected levels of demand, Gulf oil supplies have become, and are likely to remain, less important for the OECD as a whole in the 1990s than they were in the 1970s. It is on the demand, rather than the supply, side that the major uncertainty for the 1990s lies. In the absence of a major surge in oil demand, there is considerable confidence that OECD countries will be able to obtain sufficient oil at acceptable prices, even at a time of adverse political and military developments in the region. Hence there is likely to be less interest in the region *in terms of oil supply-related anxiety* in the period up to 2000.

Notes

[1] *Energy in non-OECD Countries, Selected Topics 1988*, International Energy Agency, Paris, 1988, p.27.

[2] There is a need to differentiate between major oil *producers* and major oil *exporters*. This chapter will be focusing on the latter. While additional oil production in any country will inevitably affect the volume of internationally traded oil, it is the changing contribution of the Gulf countries to international trade that we are seeking to illuminate here.

[3] The members of OPEC are: Saudi Arabia, Kuwait, United Arab Emirates, Iran, Iraq, Qatar, Algeria, Libya, Indonesia, Venezuela, Ecuador, Nigeria, Gabon.

[4] As an illustration of this point, Saudi Arabian proved reserves were revised upwards to 252 bn barrels at the beginning of 1989. Nevertheless, reserves figures issued by OPEC governments need to be treated with some caution since they are often designed to improve a country's negotiating position within the Organisation in respect of production quotas. James Tanner, 'Estimate of Saudi crude oil reserves suddenly is raised 51% by Aramco', *Wall Street Journal*, 10 January 1989.

[5] The terms: 'installed capacity', 'sustainable capacity', and 'capacity after investment' are different concepts. Installed capacity refers to the highest capacity estimate for each country; sustainable capacity refers to the largest amount of capacity that could be used continuously over a period of months; capacity after investment refers to potential production ability after the reopening of mothballed plant. Capacity after investment does not take into account possible new capacity which could be installed with the development of new reserves.

[6] Potential Saudi production capacity is the subject of endless speculation. The figures given here are an average of a large number of estimates which have been suggested to us.

[7] 'Iran and Iraq are set to rebuild their gas industries', *International Gas Report*, 2 September 1988, p.13. For pre-revolutionary plans see: Fereidun Fesharaki, *The Development of the Iranian Oil Industry: International and Domestic Aspects,* New York, Praeger, 1976, pp. 208-9.

[8] A typical example of a widely read report was: *Project Interdependence: US and World Energy Outlook Through 1990*, Senate Committee on Energy and Natural Resources and House of Representatives Committee on Interstate and Foreign Commerce, 1977. A summary report printed by the Congressional Research Service, June 1977, Publication No. 95-31, estimated (Table 20, p. 61) that, to meet demand, oil-producing countries must be able and willing to produce 64.3-87.0 mmbd by 1990, of which OPEC would be required to produce 41.4-54.2 mmbd.

[9] The export figures are for crude oil and oil products. The figures for crude oil exports only are 27.6 mmbd and 10.9 mmbd for 1977 and 1985 respectively. *OPEC Statistical Yearbook*, 1986, Tables 14 and 26, pp. 15 and 27.

[10] It is possible to come up with more startling statistics depending on one's choice of years. World primary energy consumption hit a peak in 1979 which was not surpassed again until 1983; OECD primary energy consumption declined in the period 1979-83 and rose steadily thereafter, but remained just below its 1979 level in 1987. World and OECD oil consumption peaked in 1979 and still remains substantially below those levels.

[11] For background see Paul McDonald, 'Oil and the Gulf War', *The World Today,* December 1986, pp. 202-5; reports of interviews with the Iranian oil minister and deputy oil minister can be found in *Middle East Economic Survey,* Vol. XXX, No. 39, 6 July 1987, pp. D7-8; and ibid, No. 50, 8 September 1987, pp. A2-3.

[12] See note 4; see also Frank E. Niering, 'Continued Drive Downstream', *Petroleum Economist*, April 1988, pp. 121-3.

[13] Richard Johns, 'Venezuelan oil reserves may rise by 9bn barrels', *The Financial Times,* 18 February 1988.

[14] 'Budget cuts crimp Mexico's ability to raise oil output,' *Petroleum Intelligence Weekly*, 23 May 1988, p. 5.

[15] This is a Shell UK forecast which assumes 'several years of low prices and then..a return to $25 per barrel in 1986 money...'. It is broadly consistent with the estimates adopted by the House of Commons Energy Committee in its report: *The Effect of Lower Oil and Gas Prices on Activity in the North Sea, Session 1986-87, HC 175,* 13 May 1987, pp. xxi, 78-86. The Department of Energy suggests a more optimistic scenario with oil production falling to 90-120mt (1.8-2.4 mmbd) by 1990; and to 70-105 mt (1.4-2.1 mmbd) by 1992. *Development of the Oil and Gas Resources of the United Kingdom*, HMSO, 1988, Table 8, p. 26.

[16] The Royal Ministry of Petroleum and Energy, *Fact Sheet*, Oslo, 1988, p.105.

[17] Total Soviet oil exports were 3.9 mmbd (196 mt) in 1987, comprising 1.7 mmbd to OECD countries, 2.0 mmbd to socialist countries and 0.2 mmbd to developing countries.

[18] Jonathan P. Stern, *Soviet Oil and Gas Exports to the West: Commercial Transaction or Security Threat?* Aldershot, Gower, 1987.

[19] Pessimistic analyses of Chinese oil production and demand suggest that the country may not be able to remain a net exporter in the 1990s. Edward A. Gargan, 'China looks for help to keep its oil flowing,' *New York Times,* 16 May 1988.

[20] John Cranfield, 'North Yemen: New Middle East oil exporter', *Petroleum Economist*, July 1987, pp. 257-8.

[21] The British Department of Energy annual survey: *Development of the Oil and Gas Resources of the United Kingdom*, 1988, states (p.73) that: 'The average cost per barrel for fields which started production between 1980 and 1987 is estimated to be £10 ($16)'.

[22] Much of this production decline can be attributed to stripper wells which are a unique feature of US production. 'Non-OPEC output aided by slowing US decline', *Petroleum Intelligence Weekly*, 6 June 1988, p.3; 1985-7 figures from: *IEA Quarterly Oil and Gas Statistics,* Paris, IEA/OECD, Ist Quarter 1988.

[23] Office of Technology Assessment, *US Oil Production, The Effect of Low Oil Prices*, Washington DC, September 1987, p.1.

[24] 'Oil drilling rising around the world, despite lower price,' *Petroleum Intelligence Weekly*, 18 July 1988, p.5.

[25] The study suggests that at $15 per barrel 50% of non-OPEC oil reserves could be produced; at $20 nearly 70%, and at $25 nearly 80%. Andrew B. Gordon, 'Non-OPEC oil production: potential for increasing output', *Petroleum Economist,* July 1988, pp. 234-6.

[26] Shipping attacks rose from 62 in 1984 to 187 in 1987. David Fairhall, 'War has become a nice little earner', *The Guardian,* 5 January 1988. Hamilton A. Twitchell, 'Oil and the Gulf War: In Retrospect,' *Geopolitics of Energy*, Vol. 11, No. 1, January 1989, pp. 4-6.

[27] In this context a 'sustained threat' means the ability to close the Strait to oil tanker traffic for a period exceeding one week.

[28] Quoted in Daniel Yergin and Martin Hillenbrand (eds), *Global Insecurity: A Strategy for Energy and Economic Renewal*, Boston, Mass., Houghton Miflin, 1982, p.111. A wealth of literature was devoted to the vulnerability of the Strait during the past fifteen years. See for example, R.K. Ramazani, 'Security in the Persian Gulf,' *Foreign Affairs*, Spring 1979, pp. 821-35. For details on the Strait, see R.K. Ramazani, *The Persian Gulf and the Strait of Hormuz,* Amsterdam, Sijthoff and Noordhoff, 1979.

[29] Subsequently there has been an agreement to recommence gas exports through IGAT I. 'Iran and USSR discuss using IGAT I for Oil exports', *Middle East Economic Survey*, Vol. XXX, No. 45, 17 August 1987, p.A5, 'USSR to buy 3bn cubic metres of Iranian gas', *ibid,* Vol. XXXII, No.8, 28 November 1988, p.A2.

[30] John Cranfield, 'Iran: seeking alternative export routes,' *Petroleum Economist*, May 1988, pp. 151-3.

[31] 'Iraq builds second strategic pipeline,' *Middle East Economic Survey,* 9 May 1988, Vol XXX1, No. 31, p. A5.

[32] See the evidence of witnesses from BP to the House of Commons Foreign Affairs Committee report on *Current UK Policy towards the Iran-Iraq Conflict,* 2nd Report, Session 1987-8, HC 279 (iii), paras 199-200.

[33] Paul McDonald, 'Red Sea: The Middle East's next troublespot?' *The World Today*, May 1988, pp. 76-7.

[34] For details of the Iraq-Syria pipeline see Phebe Marr, *The Modern History of Iraq*, Boulder, Westview, 1985, pp. 120, 128, 253 and 287. For the Trans-Arabian pipeline (TAPLINE) see Stockholm International Peace Research Institute, *Oil and Security*, a SIPRI Monograph, Stockholm, Almqvist and Wiksell, 1974, p.55; and David E. Long, *The United States and Saudi Arabia: Ambivalent Allies,* Boulder, Westview, 1985, pp. 15 and 21.

[35] It is interesting to speculate that, if the price collapse had occurred some years earlier, it might not have been so severe; a gradual slide might have been achieved. However, the events of 1986 appear to reinforce the view that the oil market is prone to radical price movements rather than orderly changes.

[36] 1985-7 figures from: *IEA Quarterly Oil and Gas Statistics*, Paris, IEA/OECD, Ist Quarter 1988; July 1988 figure from *Petroleum Intelligence Weekly*, 22 August 1988, p.8.

[37] United States Department of Energy, *Energy Security: A Report to the President,* Washington DC, March 1987, pp.24-5.

[38] *BP Statistical Review of World Energy*, June 1987, p.7.

[39] For example, Jose Goldenberg et al., *Energy for A Sustainable World,* Washington DC: World Resources Institute/John Wiley, 1987.

[40] A distinction needs to be drawn between the acquisition of equity for purely investment purposes, and the acquisition of assets, such as refineries and petrol retailing outlets, which are specifically aimed at assisting the marketing of an exporter's product. This distinction may be blurred in cases such as the substantial shareholding in British Petroleum by Kuwaiti interests.

[41] In April 1988, a meeting was held between OPEC representatives and the representatives of seven other exporting countries: Mexico, Colombia, Egypt, Oman, Angola, China and Malaysia (Norway and the Texas Railroad Commission sent observers). Six of the non-OPEC group (Colombia being the exception) offered to cut production by 5% for a two-month period in order to support prices, if OPEC would match this measure. Although a majority of OPEC countries backed this proposal, it was rejected primarily because of Saudi and Kuwaiti reluctance to agree. Robert J. McCartney and Martha M. Hamilton, 'Modest result foreseen from OPEC meetings', *International Herald Tribune,* 25 April 1988 and 'Six non-OPEC producers agree temporary output curbs', Ibid., 28 April 1988; the communiqué from the non-OPEC countries can be found in *Middle East Economic Survey*, Vol. XXX1, No. 30, 2 May 1988, pp. A4-6.

Table 5.1 Proven reserves of major oil producing countries (end 1987)

	Reserves (bn barrels)	% of World	Reserves to Production Ratio*
OPEC			
Gulf:			
UAE	96.2	10.7	166
Iran	92.9	10.4	111
Iraq	100.0	11.2	131
Kuwait	91.9	10.3	237
Saudi Arabia	167.0	18.6	107
Total Gulf	468.2	51.2	
Other OPEC:			
Venezuela	56.3	6.3	91
Libya	21.0	2.3	26
Algeria	8.5	0.9	25
Nigeria	16.0	1.8	34
Indonesia	8.4	0.9	18
Rest of OPEC	110.2	12.2	
Other:			
USA	33.4	3.7	9
Canada	7.7	0.9	11
Mexico	48.6	5.4	48
Norway	14.8	1.6	37
United Kingdom	5.2	0.6	6
China	18.4	2.0	18
USSR	59.0	6.6	13

* years of production remaining at 1987 levels

Source: BP Statistical Review of World Energy, June 1988, p.2.

Table 5.2 Oil production from major producers,1987(a) (mmbd)

Middle East	
Saudi Arabia	4.4
Iran	2.3
Iraq	2.1
UAE	1.6
Kuwait	1.1
Total	13.1

Africa	
Algeria	1.0
Nigeria	1.3
Libya	1.0
Total	5.2

North America	
USA	9.9
Canada	1.9
Total	11.8

Western Europe	
Norway	1.1
UK	2.6
Total	3.9

USSR and Eastern Europe	
USSR	12.5
Total	12.9

Far East/Australisia	
Indonesia	1.3
China	2.7
Total	6.1

Latin America	
Venezuela	1.6
Mexico	2.9
Total	6.6

Total World	60.2

(a) Producers of more than 1 mmbd of crude oil and NGLs

Source: Ibid.

Table 5.3 Sustainable OPEC production capacity (a) (mmbd)

	Maximum production since 1978 (b)	1987 (c)	2000 (d)
Iran	4.1	3.0	3–4
Iraq	3.7	3.0	4–4.5
Kuwait	2.6	2.0	2.4–2.6
Neutral Zone	0.6	0.6	0.4–0.5
Qatar	0.5	0.6	0.1–0.2
Saudi Arabia	10.5	8.0	9.5–11
UAE	1.9	2.2	2–2.4
Total		19.4	23.8–28.2
OPEC			27.8–33.5

(a) Sustainable capacity is the highest percentage of installed capacity that could be produced and maintained over a period of months.
(b) average daily output for one month period 1978–87
(c) production capacity in 1987
(d) projected in 1982

Sources: World Energy Outlook IEA/OECD 1982, Paris, International Energy Agency, 1982, pp. 234 and 246; Energy in non–OECD Countries, Selected Topics 1988, Paris, International Energy Agency, 1988, Table 34, p.88.

Table 5.4 Estimates of Gulf sustainable capacity, 1988 (mmbd)

	Production 1987	Sustainable Capacity A	Sustainable Capacity B
Iran	2.3	2.5–3	3–4
Iraq	2.1	3.5–4	4–5
Kuwait	1.1	2–2.5	2–2.5
Neutral Zone	0.4	0.6	0.6
Qatar	0.3	0.5	0.5
Saudi Arabia	4.0	6–7.5	8–10
UAE	1.5	2	2
Total	11.7	17.1–20.1	20.1–24.6
Spare Capacity in 1987 (capacity – production)		5.4–8.4	8.4–12.9

Capacity A = Sustainable production capacity which could be brought into operation within a matter of days/weeks, assuming available transportation capacity.

Capacity B = Sustainable production capacity after reopening mothballed plant and investing in new capacity (ie. after six months), assuming available transportation capacity.

Source: Author's estimates.

Table 5.5 OPEC crude oil and products exports (mmbd)

	1973	1977	1979	1985	1986	1987(a)
Gulf						
Iran	5.42	4.99	2.64	1.60	1.46	1.6
Iraq	1.93	2.20	3.31	1.16	1.47	1.8
Kuwait	2.85	1.94	2.51	0.97	1.25	1.2
Qatar	0.57	0.41	0.50	0.30	0.30	0.3
Saudi Arabia	7.35	8.95	9.19	2.67	4.26	3.8
UAE	1.52	1.99	1.82	1.07	1.24	1.5
Total	18.95	19.67	18.89	6.66	8.73	10.2
Non-Gulf:						
Algeria	1.03	1.07	1.03	0.54	0.58	0.8
Libya	2.21	2.03	2.05	0.99	1.17	0.9
Nigeria	1.99	2.04	2.23	1.35	1.23	1.1
Gabon	0.13	0.19	0.18	0.17	0.15	0.2
Venezuela	3.15	1.96	2.09	1.36	1.45	1.5
Ecuador	0.20	0.14	0.14	0.18	0.19	0.1
Indonesia	1.17	1.47	1.22	0.86	0.95	0.8
	8.59	7.97	7.91	4.27	4.45	5.5
OPEC(b)	29.52	29.39	28.93	13.21	15.73	15.7
Gulf OPEC as % of OPEC	66.5	69.7	69.0	58.8	63.4	65.0
Gulf as % of World trade:						
(i)	48.6	49.8	47.3	25.5	29.5	29.7
(ii)	59.7	59.5	56.2	31.3	37.4	41.6

(a) preliminary
(b) totals may not add due to rounding
(i) percentage of world trade in crude oil and products
(ii) percentage of world trade in crude oil only

Source: OPEC Annual Statistical Bulletin 1986, pp. 24,27,30 and 31.

Table 5.6 OECD energy consumption (mtoe)

	1973	%	1979	%	1987	%
Oil	1939.4	53.22	1973.6	50.8	1658.9	43.02
Gas	743.4	20.50	780.8	20.1	734.3	19.05
Coal	680.8	18.68	725.1	18.7	855.6	22.19
Hydro	235.9	6.47	268.8	6.9	280.5	7.28
Nuclear	44.4	1.22	134.5	3.5	326.0	8.46
Total	3626.6		3882.7		3855.3	

World Energy Consumption (mtoe)

	1973	%	1979	%	1987	%
Oil	2797.5	47.3	3124.5	44.9	2940.7	37.6
Gas	1066.1	18.0	1281.7	18.4	1555.8	19.9
Coal	1668.4	28.2	1968.0	28.3	2386.5	30.6
Hydro	331.5	5.6	424.5	6.1	523.9	6.7
Nuclear	49.4	0.8	154.6	2.2	404.1	5.2
Total	5913.4		6953.3		7811.0	

Source: BP Statistical Review of World Energy, 1984 and 1987.

Table 5.7 Oil production from small producers 1987 (a) (mmbd)

Net Importers		Net Exporters		Roughly Self-Sufficient	
India	0.62	Qatar	0.36(b)	Congo	0.12
Brazil	0.60	Gabon	0.16(b)	Tunisia	0.10
Australia	0.65	Ecuador	0.17(b)	Peru	0.17
Romania	0.21	Oman	0.57	Argentina	0.45
	2.08	Brunei	0.15	Syria	0.22
		Colombia	0.39		1.06
		Malaysia	0.47		
		Trinidad	0.16		
		Cameroon	0.17		
		Angola	0.34		
		Egypt	0.93		
			3.87		

(a) Producers of 0.1–0.99 mmbd of crude oil and NGLs
(b) OPEC member

Source: Energy in non–OECD Countries, Selected Topics 1988, Paris: International Energy Agency, 1988, Table 15, p.25.

Table 5.8 Oil exploration and development costs

UK		
	Pre–1980 fields	£7 [$11.00(a)]
	1980–87 fields	£10 [$16.00(a)]
(a) 1987 exchange rates		
USA		
	Stripper Wells	$7.40–9.30
	Offshore	$5.20
	Alaska North Slope	$15 inc $9 transport.
Gulf		
		$
	Iraq	0.275
	Kuwait	0.132
	Saudi Arabia	0.325
	Iran	0.586
	Qatar	1.537
	Abu Dhabi	0.984

Sources:

For the UK, Development of the Oil and Gas Resources of the United Kingdom Continental Shelf, HMSO, 1988, p. 73.

For the USA, Impact of Lower Oil Prices, Paris, IEA, April 1986, p.34.

For the Gulf, M.A.Adelman, The Competitive Floor to World Oil Prices, Mimeo, Massachusetts Institute of Technology, April 1986, p.37.

Table 5.9 Pipeline outlets from the Gulf, 1988

	Capacity (mmbd)	
1. Non-operational		
Saudi Arabia: Trans-Arabia Pipeline (TAPLINE)		
Ras Tanura-Sidon	0.5	Closed 1975
Iraq-Mediterranean		
Kirkuk-Banias/Tripoli	1.4	Closed 1982
2. Operational		
Egypt: Red Sea-Mediterranean (SUMED)	1.7	
Iraq-Turkey:		
Kirkuk-Ceyhan I	1.0	
Kirkuk-Ceyhan II	0.5	
Iraq-Petroline	0.5	
(linked to:)		
Petroline, Saudi Arabia		
Abqaiq-Yanbu	1.85	
3. Future		
Iraq IPSA II parallel to IPSA I, Saudi Petroline	1.65	to be completed in September 1989
Iran: Gurreh-Taheri	0.6-1.0	possible extension to Jask

Total pipeline capacity by end 1989:

Saudi Arabia (Petroline)	1.85
Iraq (via Turkey)	1.5
Iraq (via Saudi Arabia)	1.65(a)
	5.00

(a) assumes that existing 0.5 mmbd capacity available to Iraq via Petroline will be withdrawn once IPSA II is built.

Table 5.10 Oil-exporting countries in the 1990s

Export status in 1987 (mmbd)

Export potential by 2000 compared with 1987:	BOECs(a) (more than 1.5)	MOECs(b) (0.5–1.5)	SOECs(c) (less than 0.5)
MAJOR SPARE CAPACITY (more than 1 mmbd)	Saudi Arabia Iraq Iran UAE	Kuwait Venezuela Mexico	
MAJOR INCREASE (0.5–1.0 mmbd)		Libya Norway	
MODEST INCREASE (zero to 0.5 mmbd)			Oman, Qatar, Ecuador Gabon, Syria, Egypt Malaysia, Angola, Brunei, Colombia Trinidad, North Yemen(d)
MODEST DECLINE (zero to 0.5 mmbd)	USSR	Nigeria Algeria Indonesia	China
MAJOR DECLINE to net importer)		UK	

(a) BOECs: big oil exporting countries, (b) MOECs: medium size oil exporting countries
(c)SOECs small exporting countries, (d) negligible production in 1987

CONCLUSION

This study has argued that the Gulf area will continue to be an important sub-system of the global international community throughout the decade of the 1990s, and one which will go on preoccupying external powers. There are three main reasons for this. First, there is the massive concentration of the world's oil reserves along the shores of the Gulf. Even though, as Chapter 5 by Jonathan Stern has pointed out, Gulf oil may not have the profile of ten years ago in the world oil supply picture, the stability of the world market still necessitates the exportation of some 10 mmbd of crude oil from the Gulf. Moreover, certain states, notably Japan and Italy, that purchase a significant proportion of their oil imports from the region will follow events and trends in the area particularly closely. Ultimately, even if the Gulf oil producers do not return to centre stage until well into the next century, they continue to act as the custodians of a major energy resource. Oil importers, whether developed or developing countries, cannot treat lightly the fortunes of these custodians even when the oil market is flat. It is the stability of these states, and the region in which they exist, in the short term which will determine the ease with which oil-importing states are able to secure reliable supplies of oil when the market becomes significantly tighter.

Secondly, the Gulf will remain an area where superpower interests rub against each other. The Gulf will always be important to the Soviet Union because of its geographical proximity. The US too has major interests in the region, whether they be oil-related or linked to concern at the possible spread of ideologies that threaten to impair the Western orientation of the GCC states. Though the US will certainly reduce its naval presence in and just outside the Gulf, it is unlikely to want to give up the facilities which it has been granted as a result of its involvement in the latter stages of the Gulf conflict. Indeed, the perceived success of US naval policy in the Gulf is likely to convince defence analysts in Washington of the efficacy of the military option and the right of the US to play an active role in the affairs of the area. The new climate of detente means that there is a much reduced chance of superpower competition being translated into potential confrontation. Nevertheless, it would be naive to imagine that the constructive dialogue taking place on a strategic plane will necessarily be reproduced at a regional level. The circumstances continue to exist for competition between the superpowers for influence in the Gulf region. The state of flux which exists in the Gulf and neighbouring theatres, such as Afghanistan and the Levant, is going to increase the opportunities for the superpowers to score off one another. The attractiveness of such tactical opportunities could well prove too much to resist.

Thirdly, the power relations of the Gulf area are extremely fluid, and hence uncertainty as to the future is compounded. This will continue to be the case until a new regional political balance is achieved. During the 1970s, a pattern of political relations emerged which resulted in a high degree of stability, though the omission of Iraq from a central role in Gulf security contained the seeds of the ultimate outbreak of hostilities. The Iranian revolution upset the political balance in the northern Gulf and resulted in a long and debilitating conflict. The aftermath of the ceasefire has seen Iran and Iraq sparring with each other to formalise their maximum influence in the post-war regional sub-system of inter-state relations. It is likely to be months if not years before Baghdad and Tehran arrive at a compromise, *de facto* or *de jure*, which both are willing to accept and which defines the limits of their power in the area. Until then the boundaries of political action are likely to be clouded by bitterness, suspicion, and ambition. The grouping of the smaller Gulf states into a political collectivity has also changed the pre-war face of regional political interaction. In addition to the uncertainties of the relationship between Iran and Iraq, the way in which these two regional powers relate diplomatically to the GCC and to its individual constituents will be a process which will take a long period to identify and consolidate into stable relationships.

In spite of the uncertainties which will characterise the regional inter-relations of the Gulf, the prospects for conflict and tension appear to be low. The ending of the Gulf war was received with general relief among the populations of both belligerents. There now exists a deep popular revulsion against the resumption of any substantive military conflict involving either party. This perception of the undesirability of war and conflict has spread to the elite level. It appears that the political leaderships of both Iran and Iraq appreciate the depth of the antipathy which exists within their countries towards the resumption of warfare. Furthermore, the sudden reversal of military fortunes which characterised the eight-year war on a number of occasions has illustrated the extent to which warfare is a blunt instrument for trying to obtain a particular set of political objectives. Warfare remains a mechanism of political action which is both extremely hard to manage and which often results in hidden and negative effects. It is therefore likely that neither the authorities in Tehran nor in Baghdad will lightly dabble again in military adventures.

The end of the impetus for war on the part of the two most populous and bellicose states in the Gulf is an encouraging sign for peace. If the general short- to medium-term backdrop is one which makes peace rather than war more likely, the eradication of one of the major potential sources of tension will have done much to ensure that peace is not undermined unwittingly. It had become a cliché of strategic thinkers to cast the Strait of Hormuz as a vital but vulnerable point for Western interests, given its importance for oil exports. The diversification of oil export routes over the last decade has much reduced the Strait as a 'choke-point'. Already in excess of 4 mmbd could be exported from the Gulf area using the existing pipeline capacity and there are plans on both sides of the Gulf to expand this capacity further. A reduction in the importance of the Strait, and a change in perception to this end on the part of policy-makers, can only help lower the anxiety about the vulnerability of the main sea route. This in turn should make external actors in particular more relaxed about the security of oil supplies. Given its location in relation to the Strait, this should help to calm fears about what Iran might want or be able to do to supplies of oil. This should enable Western states to deal with Tehran in a more relaxed way.

In the main body of this study, emphasis has been put on the continuity of elites and the apparent stability of the Gulf regimes up to the end of the 1980s. It cannot automatically be assumed that this will continue into the 1990s. Many of the littoral states will probably face successions due to natural causes in the course of the next ten years. The demise of Ayatollah Khomeini in Iran, Shaikh Rashid in Dubai and Shaikh Zayid in Abu Dhabi will leave great gaps inside those polities. Indeed, these three figures are in many ways irreplaceable. In Saudi Arabia, with the three most senior princes roughly the same age and enjoying only indifferent health, the spectre of even a triple succession presents itself as a possibility in the next decade. Even if accomplished smoothly, such changes at the level of head of state, combined with the inevitable turnover in related patronage appointments, will be destabilising in themselves. Uncertainty and a preoccupation with local politics in Saudi Arabia will be particularly undesirable in the early 1990s, when the GCC will have an important role to play in re-establishing a secure and stable power equilibrium in the Gulf.

The internal problems which states of the area may face in the 1990s will not be akin to the challenges of the 1980s. The spectre of an uprising of the Shi'a communities in the Gulf states now looks unlikely to materialise. The successful blocking of the export of revolutionary Shi'ism by Iran through the medium of war has removed a potent threat to the Sunni Arab regimes of the Gulf. The threat from discontented local Shi'a groups could be replaced by a more mundane, yet ultimately possibly more real, threat: the gap between economic performance and expectation. All of the states of the Gulf are undergoing rapid expansion in their populations. The numbers of young people coming on to the labour market are likely to far outstrip the performance of the local economies. The exceptions may be Iran and Iraq, assuming that they are able to raise the necessary capital if the various grandiose visions of post-war reconstruction are fully to be realised. The continuation of a low oil price will make this more difficult. It may also accentuate the tension among the Gulf members of OPEC. The rapid expansion of the labour market may require increasingly difficult decisions. One solution to the problem has been to go for deficit budgeting, run down reserves and stomach annual current account deficits. This has been the course adopted to date by Saudi Arabia, and it has avoided a steep rise in unemployment. However, the strategy of deficit financing is a finite one. Already Riyadh's running down of its foreign reserves has compounded the revenue situation as earnings from investments have also fallen. In the case of Saudi Arabia, current spending patterns can be maintained for perhaps another five or six years without major economic problems having to be faced. This will see the kingdom through until the mid 1990s. If oil prices and consequently income continue to be flat until the end of the century stark choices may face Riyadh in the second half of the next decade.

Forecasting the future is not an exact science; indeed, it is not a science at all. One may exaggerate or underestimate particular factors which one knows will be important to the future. Then there is the problem of entirely unforeseen variables. However, the enduring impression is that the main challenge in the 1990s is likely to be to construct a new and stable system of power relations for the region - one that enables potential conflicts to be defused at an early stage, and one that protects the legitimate interests of all the Gulf states, while not threatening the interests of the various external powers. The prospect of achieving this is by no means bleak. The backdrop to the future is an intense desire for peace on the part of all the littoral states, set against no obvious prospect of major superpower conflict. The threat of attempted internal destabilisation also looks set to recede. However, hard choices dictated by the need to allocate increasingly finite financial resources may cause the sort of political problems which the ideologically-motivated movements so palpably failed to stir up in the preceding decades.

Index